EURO Advanced Tutorials on Operational Research

Series editors

M. Grazia Speranza, Brescia, Italy
José Fernando Oliveira, Porto, Portugal

More information about this series at http://www.springer.com/series/13840

Cláudio Alves • François Clautiaux •
José Valério de Carvalho • Jürgen Rietz

Dual-Feasible Functions for Integer Programming and Combinatorial Optimization

Basics, Extensions and Applications

Cláudio Alves
Department of Production and Systems
University of Minho
Braga, Portugal

François Clautiaux
Institut de Mathématiques de Bordeaux
University of Bordeaux
Talence, France

José Valério de Carvalho
Department of Production and Systems
University of Minho
Braga, Portugal

Jürgen Rietz
Centro Algoritmi
University of Minho
Braga, Portugal

ISSN 2364-687X   ISSN 2364-6888   (electronic)
EURO Advanced Tutorials on Operational Research
ISBN 978-3-319-27602-1   ISBN 978-3-319-27604-5   (eBook)
DOI 10.1007/978-3-319-27604-5

Library of Congress Control Number: 2015960796

Springer Cham Heidelberg New York Dordrecht London
© Springer International Publishing Switzerland 2016
This work is subject to copyright. All rights are reserved by the Publisher, whether the whole or part of the material is concerned, specifically the rights of translation, reprinting, reuse of illustrations, recitation, broadcasting, reproduction on microfilms or in any other physical way, and transmission or information storage and retrieval, electronic adaptation, computer software, or by similar or dissimilar methodology now known or hereafter developed.
The use of general descriptive names, registered names, trademarks, service marks, etc. in this publication does not imply, even in the absence of a specific statement, that such names are exempt from the relevant protective laws and regulations and therefore free for general use.
The publisher, the authors and the editors are safe to assume that the advice and information in this book are believed to be true and accurate at the date of publication. Neither the publisher nor the authors or the editors give a warranty, express or implied, with respect to the material contained herein or for any errors or omissions that may have been made.

Printed on acid-free paper

Springer International Publishing AG Switzerland is part of Springer Science+Business Media (www.springer.com)

Preface

The concept of dual-feasible function (DFF) has been used to improve the resolution of several combinatorial optimization problems involving knapsack inequalities like cutting and packing, scheduling, and vehicle routing problems. DFF were used for the first time by Lueker (1983) to obtain lower bounds for the bin-packing problem. Since then, the main application of DFF was in the computation of lower bounds, even though other applications do exist, as, for instance, the generation of valid inequalities for integer programs (Chvátal 1973).

During many years, DFF were seen only as mere rounding functions that lead to lower bounds for standard packing problems by changing the value of the input data. In this tutorial, we bring a broader perspective to the subject by discussing it within the general framework of duality. A revision of the standard concepts, properties, and instances is provided with illustrative examples. We show that many lower bounds derived for packing problems can be expressed as DFF. Also we explore relevant extensions of standard DFF and their application to different combinatorial optimization problems.

The link between DFF, column generation models, and the underlying Dantzig-Wolfe decomposition is strong. The classical DFF rely on the dual perspective of the well-known column generation model of Gilmore and Gomory for the cutting stock problem. Many functions were proposed within this specific context. We explore the general properties that identify the best DFF. Additionally, we describe the general approaches that can be followed to derive new non-dominated functions. In particular, it is shown how to derive high-quality DFF from superadditive functions using symmetry. We show how to use them to derive concrete examples, and we further illustrate these ideas through the analysis of some of the current DFF that lead to the best results reported in the literature.

A first generalization of the classical DFF can be done by considering the general formulation of a set covering model somehow disconnected from the cutting stock problem. Here, we analyze this generalization and explore how it can influence (or act upon) the general properties of classical DFF. Extending DFF to nonclassical formulations is usually not straightforward. In this tutorial, we show how this extension can be done for different cases including the existence of two-dimensional rectangular items, conflicts, and general domains. Examples are provided to guide the readers through the rationale behind the development of these extensions and to provide the basis for future contributions.

Extensions and applications within the scope of integer and linear programming will be also discussed. Recent developments extending DFF to the domain of negative values are described. These generalized DFF can be used to derive valid inequalities for integer programs from a single constraint or a set of generating constraints.

This monograph was primarily written for graduate students and also advanced undergraduate students in operations research/management science, computer science, and mathematics. It requires a knowledge of linear and integer programming and duality theory. Chapters 1 and 2 present the basic concepts and definitions of DFF. Chapter 3 contains extensions of DFF to wider domains. Chapters 4 and 5 address applications in cutting and packing problems and in deriving valid inequalities, respectively. Exercises are also proposed at the end of each chapter covering the different parts of the material. Solutions or hints to a selected set of exercises are provided at the end of the book.

The authors have been using DFF in research for many years, and these proved to be very useful for the efficient computation of lower bounds for many different integer programs and combinatorial optimization problems, with nice computational results. Additionally, they also used them to derive valid inequalities. DFF are problem dependent. We hope that the insight provided by this monograph may foster research and a more widespread use of DFF in other problems. The future still holds much to be discovered in this area.

We would like to thank many colleagues and researchers, who discussed many issues and stimulated our interest and research in this area. The authors also would like to thank Gleb Belov, Julia Bennell, Rita Macedo, and Daniel Porumbel, who did a careful review of an earlier draft of this manuscript and helped to improve its quality.

This work was supported by FEDER funding through the Programa Operacional Factores de Competitividade (COMPETE) and by national funding through the Portuguese Science and Technology Foundation (FCT) in the scope of the project PTDC/EGE-GES/116676/2010 (Reference from COMPETE: FCOMP-01-0124-FEDER-020430), and by FCT within the project scope UID/CEC/00319/2013.

Braga, Portugal	Cláudio Alves
Talence, France	François Clautiaux
Braga, Portugal	José Valério de Carvalho
Braga, Portugal	Jürgen Rietz
October 2015	

Contents

1 Linear and Integer Programming 1
   1.1 Introduction 1
   1.2 Dantzig-Wolfe Decomposition 3
       1.2.1 Reformulation of the Original Model 3
       1.2.2 Dantzig-Wolfe Decomposition in Integer Programming 4
   1.3 Structure of DW-Decomposition Models 6
       1.3.1 Gilmore and Gomory Model for the Cutting Stock Problem 7
       1.3.2 Block Angular Structure 8
       1.3.3 Parallel Non-identical Machine Scheduling 9
       1.3.4 Solution of DW-Models with Column Generation 10
   1.4 Duality and Bounds from Dual Feasible Solutions 11
   1.5 Examples 12
       1.5.1 One-Dimensional Cutting Stock Problem 12
       1.5.2 Vector Packing Problem 14
   1.6 Related Literature 16
   1.7 Exercises 16

2 Classical Dual-Feasible Functions 21
   2.1 Introduction 21
   2.2 Properties 24
       2.2.1 Maximality 24
       2.2.2 Maximality of Convex Functions 27
       2.2.3 Extremality 27
       2.2.4 Extremality of Convex Functions 29
   2.3 Generating One-Dimensional Dual-Feasible Functions 33
       2.3.1 Linear Combination 33
       2.3.2 Composition 34
       2.3.3 Symmetry 35
       2.3.4 Using the Limiting Behaviour of a Function 36
       2.3.5 Rounding Functions 37

2.4	Examples		39
	2.4.1	Applying Symmetry	39
	2.4.2	Using Rounding Functions and Applying Symmetry	39
	2.4.3	Improving a Function by Using Its Limiting Behaviour	41
	2.4.4	A Special Case: A Staircase Function with Infinitely Many Stairs	43
2.5	Related Literature		45
2.6	Exercises		46

3 General Dual-Feasible Functions ... 51

3.1	Introduction		51
3.2	Extension of Dual-Feasible Functions to General Domains		52
	3.2.1	Definition	52
	3.2.2	Maximality	54
	3.2.3	Extremality	56
3.3	Applications		58
3.4	Properties of Maximal General Dual-Feasible Functions		60
	3.4.1	Structure	61
	3.4.2	Behaviour at Given Points	64
	3.4.3	Limits of Possible Convexity	66
	3.4.4	Composition and Convex Combinations	66
3.5	Examples		67
3.6	Building Maximal General Dual-Feasible Functions		73
	3.6.1	Method I	73
	3.6.2	Method II	76
	3.6.3	Method III	77
	3.6.4	Examples	80
3.7	Related Literature		86
3.8	Exercises		87

4 Applications for Cutting and Packing Problems ... 91

4.1	Introduction		91
4.2	Set-Covering Dual-Feasible Functions		91
	4.2.1	Data-Dependent Dual-Feasible Functions	93
	4.2.2	Data-Independent Dual-Feasible Functions	94
	4.2.3	General Properties	94
4.3	Vector Packing Dual-Feasible Functions		95
	4.3.1	Basic Definition	95
	4.3.2	General Properties of VP-MDFF	98
	4.3.3	General Classes of VP-MDFF	102
4.4	Orthogonal Packing		110
	4.4.1	DFF for the Oriented Case (m-OPP-O-DFF)	111
	4.4.2	DFF for the Case with Rotation (m-OPP-R-DFF)	112
4.5	Bin-Packing		113

	4.6	Bin-Packing Problem with Conflicts	115
		4.6.1 BPC-DDFF Based on a Knapsack Subproblem	115
		4.6.2 A BPC-DDFF Based on Graph Decomposition	117
	4.7	Related Literature	119
	4.8	Exercises..	120
5	**Other Applications in General Integer Programming**		125
	5.1	Superadditive Functions in Integer Programming	125
	5.2	Valid Inequalities for Integer Programs	126
	5.3	Examples	127
	5.4	Related Literature	130
	5.5	Exercises..	131

Appendix A   Hints and Solutions to Selected Exercises.................... 133

References... 157

Index 159

Acronyms

1D-CSP	One-dimensional cutting stock problem
BP	Bin-packing problem
BP-DDFF	Bin-packing data-dependent dual-feasible function
BPC	Bin-packing problem with conflicts
BPC-DDFF	Bin-packing problem with conflicts data-dependent dual-feasible function
CS-DFF	Cutting stock dual-feasible function
CS-MDFF	Maximal cutting stock dual-feasible function
DFF	Dual-feasible function
DW	Dantzig-Wolfe
EMDFF	Extreme maximal dual-feasible function
IP	Integer programming
KP-01	Binary knapsack problem
KPC	Binary knapsack problem with conflicts
LP	Linear programming
MDFF	Maximal dual-feasible function
mD-KP	m-dimensional knapsack problem
mD-VPP	m-Dimensional vector packing problem
m-OPP	m-Dimensional orthogonal bin-packing problem
m-OPP-O	m-Dimensional orthogonal bin packing problem with fixed orientation
m-OPP-O-DFF	m-Dimensional orthogonal bin-packing with fixed orientation dual-feasible function
m-OPP-R	m-Dimensional orthogonal bin-packing problem with rotation
m-OPP-R-DFF	m-Dimensional orthogonal bin packing with rotation dual-feasible function
SC-DFF	Set covering dual-feasible function
SC-DDFF	Data-dependent set covering dual-feasible function
VP-DFF	Vector packing dual-feasible function
VP-MDFF	Maximal vector packing dual-feasible function

Chapter 1
Linear and Integer Programming

1.1 Introduction

Integer Programming (IP) is a modelling tool that has been widely applied in the last decades to obtain solutions for complex real problems, as those that arise in cutting and packing, location, routing and many other areas. IP models are of the form:

$$\begin{aligned} \min \ & z_{IP} := \mathbf{c}^\top \mathbf{x} \\ \text{s. to} \ & \mathbf{Ax} \geq \mathbf{b} \\ & \mathbf{x} \geq \mathbf{o} \text{ and integer,} \end{aligned} \quad (1.1)$$

where $\mathbf{A} \in \mathbb{R}^{m \times n}$, $\mathbf{b} \in \mathbb{R}^m$, $\mathbf{c} \in \mathbb{R}^n$, and $\mathbf{x} \in \mathbb{Z}_+^n$ is a vector of decision variables. The zero vector is denoted by $\mathbf{o}$. The set of solutions of the IP model is $X_{IP} = \{\mathbf{x} \in \mathbb{Z}_+^n : \mathbf{Ax} \geq \mathbf{b}\}$. If the problem is feasible, the set of solutions is a discrete set, either finite or countably infinite.

A related problem that plays a central role in the resolution of an IP model is the Linear Programming (LP) model that results from relaxing the *integrality constraints*, which impose that the decision variables $\mathbf{x}$ can only take integer values. The LP model that accepts decision variables $\mathbf{x} \in \mathbb{R}_+^n$ is as follows:

$$\begin{aligned} \min \ & z_{LP} := \mathbf{c}^\top \mathbf{x} \\ \text{s. to} \ & \mathbf{Ax} \geq \mathbf{b} \\ & \mathbf{x} \geq \mathbf{o}. \end{aligned} \quad (1.2)$$

The set of solutions of the *LP relaxation* (or *continuous relaxation*) of the IP model is a convex set $X_{LP} = \{\mathbf{x} \in \mathbb{R}_+^n : \mathbf{Ax} \geq \mathbf{b}\}$. The set X_{LP} contains all the integer solutions, as $X_{IP} = X_{LP} \cap \mathbb{Z}_+^n$.

© Springer International Publishing Switzerland 2016
C. Alves et al., *Dual-Feasible Functions for Integer Programming and Combinatorial Optimization*, EURO Advanced Tutorials on Operational Research, DOI 10.1007/978-3-319-27604-5_1

Linear Programming Duality theory is a tool to derive bounds to the values of the optimal solutions of LP models. For instance, given a minimization problem, a solution that is feasible to the dual model provides a lower bound to the value of the minimum. Bounds of good quality are often used in optimization, because they can improve substantially the performance of algorithms. As we will discuss, the feasible solutions of the dual problems of the LP relaxations of some models may provide better lower bounds than others.

Typically there are different ways of modelling an IP problem, and any model is valid if its set of feasible integer solutions is X_{IP}. However, there may be differences in their LP relaxations. Some models may provide a closer description of the set of the feasible integer solutions. Let $X_1 \subseteq \mathbb{R}^n$ and $X_2 \subseteq \mathbb{R}^n$ be the convex sets of solutions of the LP relaxations of two valid IP models, meaning that $X_{IP} = X_1 \cap \mathbb{Z}_+^n = X_2 \cap \mathbb{Z}_+^n$. A model with a set X_1 is said to be stronger than a model with a set X_2, if $X_1 \subset X_2$. Alternatively one may state that the second model is weaker.

The convex hull of a set of points is the smallest convex set that contains all the points in the set. The convex hull of X_{IP}, denoted as $\text{Conv}\{X_{IP}\}$, is the strongest model for an IP problem. Using LP, the optimal solution would never be a fractional solution, because its extreme points are all integer. The issue is that it may even be difficult to know all the constraints that are needed to define the set $\text{Conv}\{X_{IP}\}$. One has to resort to other alternatives that are weaker than $\text{Conv}\{X_{IP}\}$, but may be stronger than the LP relaxation.

One technique to obtain strong models is the Dantzig-Wolfe (DW) decomposition. One drawback is that the models that result from DW-decomposition typically have an exponential number of decision variables, and it is not practical to enumerate them all. One may resort to column generation algorithms, which have been successfully used in many IP problems, but they are complex, time consuming to implement and may still represent a heavy computational burden.

We aim at implementing algorithms that use feasible solutions of the dual polyhedra of the (strong) column generation models, without solving the column generation model or enumerating any column. We derive the dual feasible solutions from the structure of the columns, and so they are valid for any instance of a given problem.

The motivation is the following. Dual Feasible Functions (DFF) provide feasible dual solutions of strong models whose corresponding lower bounds are often very close to the lower bounds provided by column generation models. As DFF can be computed quickly, the computational burden can be small when compared to column generation algorithms. Furthermore, for a given problem, it is often possible to derive not only a single DFF but several DFF, or even families of different DFF, each providing a different dual feasible solution. By using several DFF, we aim at obtaining at least a feasible dual solution that provides a good lower bound.

This chapter is structured as follows. In the following section, we present the DW-decomposition, and illustrate how it can lead to a strong model. Then, we focus on the structure of the models that result from DW-decomposition, which provides insight on the space of dual feasible solutions. Finally, two small examples are presented, showing that dual feasible solutions may provide good quality bounds.

1.2 Dantzig-Wolfe Decomposition

1.2.1 *Reformulation of the Original Model*

DW-decomposition may be applied to LP models whose constraints can be divided into two sets: the first set includes general constraints $\mathbf{Ax} \geq \mathbf{b}$, while the second set has constraints with a special structure that define a set denoted by X:

$$\begin{aligned} \min \ z_{LP} &:= \mathbf{c}^\top \mathbf{x} \\ \text{s. to} \ \mathbf{Ax} &\geq \mathbf{b} \\ \mathbf{x} &\in X \\ \mathbf{x} &\geq \mathbf{0}. \end{aligned} \quad (1.3)$$

Minkowski's theorem states that any point $\mathbf{x}$ of a non-empty polyhedron X can be expressed as a convex combination of the extreme points of X plus a non-negative combination of the extreme rays of X. Therefore, the set X can be defined as follows:

$$X = \left\{ \mathbf{x} \in \mathbb{R}_+^n : \mathbf{x} = \sum_{p \in P} \lambda_p \mathbf{x}_p + \sum_{r \in R} \mu_r \mathbf{y}_r, \sum_{p \in P} \lambda_p = 1, \lambda_p \geq 0, \forall p \in P, \right. \\ \left. \mu_r \geq 0, \forall r \in R \right\},$$

where $P = \{\mathbf{x}_p\}$ is the set of extreme points of X and $R = \{\mathbf{y}_r\}$ is the set of extreme rays of X. Replacing the value of $\mathbf{x}$ in the original model, and rearranging the terms, we obtain the following reformulation (or *DW-model*) of the problem:

$$\begin{aligned} \min \ z_{DW} &:= \sum_{p \in P} (\mathbf{c}^\top \mathbf{x}_p) \lambda_p + \sum_{r \in R} (\mathbf{c}^\top \mathbf{y}_r) \mu_r \\ \text{s. to} \ &\sum_{p \in P} \lambda_p (\mathbf{Ax}_p) + \sum_{r \in R} \mu_r (\mathbf{Ay}_r) \geq \mathbf{b} \\ &\sum_{p \in P} \lambda_p = 1 \\ &\lambda_p \geq 0, \forall p \in P \\ &\mu_r \geq 0, \forall r \in R. \end{aligned}$$

The decision variables of the reformulated problem are the variables λ_p and μ_r. The elements $\mathbf{c}^\top \mathbf{x}_p$ and $\mathbf{c}^\top \mathbf{y}_r$ define the objective function coefficients, while the columns $\mathbf{Ax}_p$ and $\mathbf{Ay}_r$ define the constraints of the reformulated problem.

The constraint on the sum of the values of variables λ_p is usually referred to as the convexity constraint. This DW-model is equivalent to the original model, as the definition of **x** provides the correspondence between any solution of the DW-model to a solution of the original model.

1.2.2 Dantzig-Wolfe Decomposition in Integer Programming

A motivation for using DW-decomposition in IP is to obtain a model stronger than the LP relaxation, which is possible when the set X has special characteristics. DW-decomposition provides a strong model if the polyhedron X does not have the *integrality property*, meaning that its extreme points and rays are not all integer. To obtain a strong model, we use $\mathbf{x} \in \text{Conv}\{\mathbf{x} \in X \text{ and integer}\} \subseteq X$, instead of using $\mathbf{x} \in X$. This corresponds to imposing the integrality constraints on the decision variables **x** only in the second set of constraints of the original model, and using integer extreme points and rays.

The set of feasible solutions of the reformulated model, expressed in terms of the variables **x** of the original model, is $X_{DWI} = \{\mathbf{x} \in \text{Conv}\{\mathbf{x} \in X \text{ and integer}\} : \mathbf{Ax} \geq \mathbf{b}, \mathbf{x} \geq \mathbf{o}\}$. The corresponding DWI-model is as follows:

$$
\begin{aligned}
\min \quad & z_{DWI} := \mathbf{c}^\top \mathbf{x} \\
\text{s. to} \quad & \mathbf{Ax} \geq \mathbf{b} \\
& \mathbf{x} \in \text{Conv}\{\mathbf{x} \in X \text{ and integer}\} \\
& \mathbf{x} \geq \mathbf{o}.
\end{aligned}
\quad (1.4)
$$

Given that $\text{Conv}\{X_{IP}\} \subseteq X_{DWI} \subseteq X_{LP}$, it follows that $z_{IP} \leq z_{DWI} \leq z_{LP}$ for minimization problems. Note that, if the polyhedron X has the integrality property, then $\text{Conv}\{\mathbf{x} \in X \text{ and integer}\} = X$, and $X_{DWI} = \{\mathbf{x} \in X : \mathbf{Ax} \geq \mathbf{b}, \mathbf{x} \geq \mathbf{o}\}$. In this case, the DWI-model is as strong as the LP relaxation, and $z_{IP} \leq z_{DWI} = z_{LP}$.

Example 1.1 The comparison of the sets X_{LP} and X_{DWI} in Fig. 1.1 illustrates how DW-decomposition may provide a strong model. The domain of the IP problem is the finite set of full dots that belong to the space of the LP relaxation, X_{LP}, shown in Fig. 1.1a, delimited by the double line.

The set of constraints is decomposed into the first set of general constraints, which is the single constraint $\mathbf{A}^1\mathbf{x} \leq b^1$, and the second set, which is the single constraint $\mathbf{A}^2\mathbf{x} \leq b^2$ and the non-negativity constraints $\mathbf{x} \geq \mathbf{o}$, which defines the set X. The set X does not have the integrality property, because it has a fractional extreme point in the x_1 axis. The set $\text{Conv}\{\mathbf{x} \in X \text{ and integer}\}$ is delimited by the double line, in Fig. 1.1b, and its extreme points are all integer.

The solution space of the reformulated model, X_{DWI}, is shown in Fig. 1.1c, delimited by the double line. It is the set of points $\mathbf{x} \in \text{Conv}\{\mathbf{A}^2\mathbf{x} \leq b^2, \mathbf{x} \geq \mathbf{o} \text{ and integer}\}$ that also obey the constraint in the first set. □

1.2 Dantzig-Wolfe Decomposition

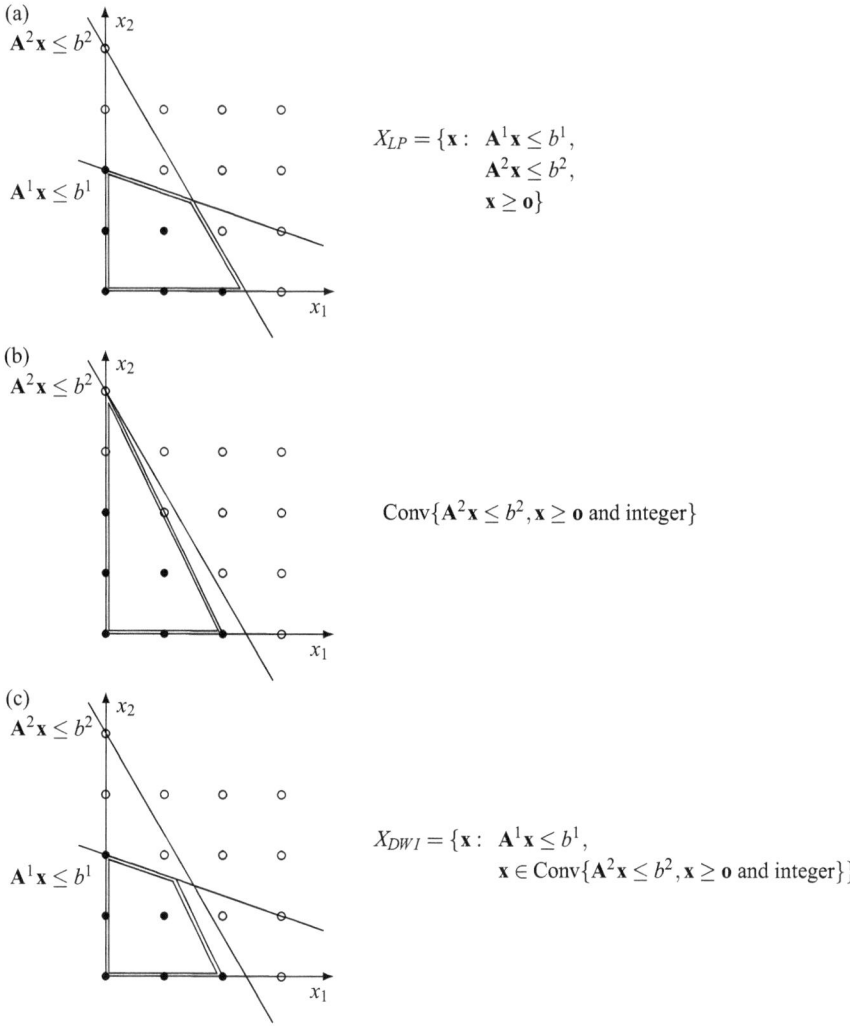

Fig. 1.1 Getting a strong model with Dantzig-Wolfe decomposition. (**a**) The set $X_{LP} = \{\mathbf{x} : \mathbf{A}^1\mathbf{x} \leq b^1, \mathbf{A}^2\mathbf{x} \leq b^2, \mathbf{x} \geq \mathbf{o}\}$. (**b**) The set Conv$\mathbf{A}^2\mathbf{x} \leq b^2, \mathbf{x} \geq \mathbf{o}$ and integer}. (**c**) The set $X_{\text{DWI}} = \{\mathbf{x} : \mathbf{A}^1\mathbf{x} \leq b^1, \mathbf{x} \in \text{Conv}\{\mathbf{A}^2\mathbf{x} \leq b^2, \mathbf{x} \geq \mathbf{o} \text{ and integer}\}\}$

Example 1.2 The IP model in Fig. 1.1 is max$\{x_1 + x_2 : x_1 + 3x_2 \leq 6, 8x_1 + 5x_2 \leq 20, x_1, x_2 \geq 0 \text{ and integer}\}$. Note that this is a maximization problem with $\leq$ constraints, but the same reformulation principles do apply. Keeping the single constraint $x_1 + 3x_2 \leq 6$ in the master problem and letting the set $X := \{\mathbf{x} : 8x_1 + 5x_2 \leq 20, x_1, x_2 \geq 0\}$, the matrix $\mathbf{A}$ is just the vector $(1, 3)$ and the vector $\mathbf{c} = (1, 1)^\top$. The extreme points of Conv$\{\mathbf{x} \in X \text{ and integer}\}$

are $\mathbf{x}_1 = (0,0)^\top, \mathbf{x}_2 = (2,0)^\top$ and $\mathbf{x}_3 = (0,4)^\top$, and there are no extreme rays, because X is a bounded set. Therefore, $[\mathbf{c}^\top\mathbf{x}_1, \mathbf{c}^\top\mathbf{x}_2, \mathbf{c}^\top\mathbf{x}_3] = [0,2,4]$ and $[\mathbf{A}\mathbf{x}_1, \mathbf{A}\mathbf{x}_2, \mathbf{A}\mathbf{x}_3] = [0,2,12]$. The DWI-model resulting from this decomposition is:

$$\begin{aligned}
\max z_{DWI} &:= 0\lambda_1 + 2\lambda_2 + 4\lambda_3, \\
\text{s. to} \quad & 0\lambda_1 + 2\lambda_2 + 12\lambda_3 \leq 6, \\
& \lambda_1 + \lambda_2 + \lambda_3 = 1, \\
& \lambda_1, \lambda_2, \lambda_3 \geq 0.
\end{aligned}$$

The $(\lambda_1, \lambda_2, \lambda_3)$ coordinates of the extreme points of X_{DWI} are $(1,0,0)$, $(0,1,0)$, $(\frac{1}{2}, 0, \frac{1}{2})$ and $(0, \frac{3}{5}, \frac{2}{5})$, respectively. The last extreme point, which is the optimal solution, maps to the solution in the original space $\mathbf{x}^* = 0\mathbf{x}_1 + \frac{3}{5}\mathbf{x}_2 + \frac{2}{5}\mathbf{x}_3 = (\frac{6}{5}, \frac{8}{5})^\top$. □

1.3 Structure of DW-Decomposition Models

Many integer programming and combinatorial optimization problems were derived using DW-decomposition, leading to reformulated models with columns that have a special structure. A column may represent a cutting pattern in a cutting stock problem, a plan for a machine in a machine scheduling problem or a trip of a vehicle in a vehicle routing problem, and the feasible solutions of the models are built combining a subset of columns. We focus on the structure of the columns, because it determines the structure of the constraints of their corresponding dual models.

There are many examples of models that were developed without any reference to DW-decomposition. They were naturally derived, because their feasible solutions can be expressed as a combination of columns with some special structure. Nevertheless, we should always keep in mind the DW-decomposition, because the integrality property helps in recognizing when those "column models" are weak or strong.

In the following sections, we present the Gilmore and Gomory model for the Cutting Stock Problem, which is the model underlying the classical dual-feasible functions addressed in Chap. 2, and the application of DW-decomposition to models with a block angular structure, which provides insight on how to derive dual-feasible functions for more general domains. Finally, we briefly describe column generation algorithms.

1.3.1 Gilmore and Gomory Model for the Cutting Stock Problem

Given a set of $m \in \mathbb{N}$ different item lengths ℓ_i ($i \in \{1, \ldots, m\}$) to be cut from stock rolls of length $L > 0$, the 1-dimensional cutting stock problem (1D-CSP) consists of finding how to cut these items such that the number of used rolls is minimized. The Gilmore and Gomory model for the Cutting Stock Problem is as follows:

$$\min \ z := \sum_{j \in J} x_j \tag{1.5}$$

$$\text{s. to} \ \mathbf{A}\mathbf{x} \geq \mathbf{b}, \tag{1.6}$$

$$x_j \in \mathbb{N}, \ \forall j \in J, \tag{1.7}$$

where $\mathbf{b} \in \mathbb{N}^m$ describes the order demands of the items, and J is the index set of all feasible patterns $\mathbf{a}^j$, which form the matrix $\mathbf{A}$. The quantity of the j-th pattern to be used is x_j. A pattern $\mathbf{a}^j \in \mathbb{N}^m$ is feasible if and only if

$$\mathbf{l}^\top \mathbf{a}^j \leq L, \tag{1.8}$$

i.e., the total sum of the lengths of all the items to be cut in the corresponding quantity does not exceed the length of the roll L.

The dual of the continuous relaxation of (1.5)–(1.7) is

$$\max \ z^D := \mathbf{b}^\top \mathbf{u} \tag{1.9}$$

$$\text{s. to} \ \mathbf{u}^\top \mathbf{a}^j \leq 1, \ \forall j \in J, \tag{1.10}$$

$$\mathbf{u} \in \mathbb{R}_+^m. \tag{1.11}$$

Both the primal and the dual problem are solvable, if no item is longer than the length of the stock rolls. Due to the strong duality theorem, the optimal objective function values of the continuous relaxation of (1.5)–(1.7) and its dual (1.9)–(1.11) are the same, and for all feasible solutions of both problems, one has $z \geq z^D$. Note that because of the constraints (1.10) and (1.11), it follows that $u_i \leq 1$ for all i.

The solutions of the Gilmore and Gomory model result from a non-negative combination of columns. It can be shown that this structure results from a DW-decomposition of an original arc-flow model, whose solutions are extreme rays that can be associated to solutions of a knapsack problem (see Sects. 1.6 and 1.7). Therefore, the convexity constraint is not needed.

Example 1.3 Consider a 1D-CSP instance with rolls of length 8 and items of lengths 4, 3 and 2, with order demands of 5, 4 and 8, respectively. A model is as follows:

	Cutting patterns						Demand b_i
$L = 8$	x_1	x_2	x_3	x_4	x_5	x_6	
$\ell_i = 4$	2	1	1				≥ 5
3		1		2	1		≥ 4
2			2	1	2	4	≥ 8
min	1	1	1	1	1	1	

Each column describes the number of items of each length produced in a cutting pattern. A feasible cutting plan is a combination of cutting patterns that satisfies the demand. The objective is to minimize the number of rolls used.

The structure of the model is appealing. It is a covering model, in which one has to select a set of columns that cover the demand. If the values of the demands were all equal to one, clearly the feasible cutting patterns should only have one item of each type. In this special case, the model would be a set covering model. □

1.3.2 Block Angular Structure

Consider a vector of decision variables $\mathbf{x}$ partitioned in $\mathbf{x} = [\mathbf{x}^1, \ldots, \mathbf{x}^i, \ldots, \mathbf{x}^K]$, with $\mathbf{x}^i \in \mathbb{R}_+^{n_i}$, $n_i \in \mathbb{N}$, $i = 1, \ldots, K$. There are K sets of constraints that define independent sets X^i, $i = 1, 2, \ldots, K$, and a set of general linking constraints. The model has a *block angular structure* and is as follows:

$$\begin{aligned}
\min z_{LP} := \mathbf{c}^{1\top}\mathbf{x}^1 &+ \mathbf{c}^{2\top}\mathbf{x}^2 + \ldots + \mathbf{c}^{K\top}\mathbf{x}^K \\
\text{s. to} \quad \mathbf{A}^1\mathbf{x}^1 &+ \mathbf{A}^2\mathbf{x}^2 + \ldots + \mathbf{A}^K\mathbf{x}^K \geq \mathbf{b} \\
\mathbf{x}^1 &\in X^1 \\
\mathbf{x}^2 &\in X^2 \\
&\ldots \\
\mathbf{x}^K &\in X^K \\
\mathbf{x}^i \in \mathbb{R}_+^{n_i}, \; n_i &\in \mathbb{N}, \; i = 1, 2, \ldots, K,
\end{aligned}$$

where $\mathbf{A}^i \in \mathbb{R}^{m \times n_i}$, $\mathbf{c}^i \in \mathbb{R}^{n_i}$, with $n_i \in \mathbb{N}$, $i = 1, 2, \ldots, K$, and $\mathbf{b} \in \mathbb{R}^m$.

In this case, consider that the sets X^i are bounded, and so there are no extreme rays. The set X^i can be defined as follows:

$$X^i = \{\mathbf{x}^i \in \mathbb{R}_+^{n_i} : \mathbf{x}^i = \sum_{p \in P^i} \lambda_p^i \mathbf{x}_p^i, \sum_{p \in P^i} \lambda_p = 1, \lambda_p^i \geq 0, \forall p \in P^i\}, \; i = 1, 2, \ldots, K,$$

1.3 Structure of DW-Decomposition Models

where $P^i = \{\mathbf{x}_p^i\}$ is the set of extreme points of X^i. The reformulation is as follows:

$$\min z_{DW} := \sum_{p \in P^1} (\mathbf{c}^{1\top}\mathbf{x}_p^1)\lambda_p^1 + \sum_{p \in P^2} (\mathbf{c}^{2\top}\mathbf{x}_p^2)\lambda_p^2 + \ldots + \sum_{p \in P^K} (\mathbf{c}^{K\top}\mathbf{x}_p^K)\lambda_p^K$$

$$\text{s. to} \quad \sum_{p \in P^1} \lambda_p^1(\mathbf{A}^1\mathbf{x}_p^1) + \sum_{p \in P^2} \lambda_p^2(\mathbf{A}^2\mathbf{x}_p^2) + \ldots + \sum_{p \in P^K} \lambda_p^K(\mathbf{A}^K\mathbf{x}_p^K) \geq \mathbf{b}$$

$$\sum_{p \in P^1} \lambda_p^1 = 1$$

$$\sum_{p \in P^2} \lambda_p^2 = 1$$

$$\ldots$$

$$\sum_{p \in P^K} \lambda_p^K = 1$$

$$\lambda_p^i \geq 0, \forall p \in P^i, \, i = 1, 2, \ldots, K.$$

In this reformulated model, a column may still represent a cutting pattern or a plan for a machine. Nevertheless, this more general structure allows us to model the cases in which the cutting patterns come from rolls of different sizes, as happens in the multiple roll lengths cutting stock problem, or the machines have different capabilities or availabilities, as in the parallel non-identical machine scheduling problem. Even though the columns are generated from different entities, their contributions are used to build feasible solutions that must obey the linking constraints.

1.3.3 Parallel Non-identical Machine Scheduling

Consider the problem of building a feasible schedule for a set of parallel non-identical machines. A plan for a machine is feasible if the jobs assigned to the machine can be executed within a given time slot. There is a set of feasible plans, one for each different machine. The plan for each machine is one selected from several plans. Each plan for a given machine represents an extreme point of the solution set of the machine. Therefore, a convexity constraint is needed to select just one plan for each machine.

The model that results from DW-decomposition is a set partitioning model (the derivation is left as an exercise). The columns represent the feasible plans for the machines. The model has a set of constraints for the jobs, which indicate that each job is executed once, and a set of constraints for the machines, which are convexity constraints.

Example 1.4 Consider an example with four jobs and two machines. Let y_k^i be a decision variable that represents a feasible plan, indexed by k, that assigns a set of jobs to machine i. Each feasible plan for a machine has a cost, and the objective is to minimize the total cost.

	y_0^1	y_1^1	y_2^1	y_3^1	y_4^1	y_5^1	y_0^2	y_1^2	y_2^2	y_3^2	y_4^2	y_5^2	
Job 1		1	1	1				1	1	1			= 1
2		1			1	1		1			1		= 1
3			1		1				1			1	= 1
4				1		1				1	1	1	= 1
Machine 1	1	1	1	1	1	1							= 1
2							1	1	1	1	1	1	= 1
min	0	27	25	24	22	21	0	12	16	14	10	14	

Note that each machine has a schedule that is the null solution, meaning that the machine is idle in the plan. Clearly, the machine constraints might be replaced by $\leq$ constraints, because the idle machine columns are slack variables for those constraints. □

1.3.4 Solution of DW-Models with Column Generation

The DWI-model may be stronger, but it comes at a price. The first issue is that there is generally an exponential number of extreme points and extreme rays in the set Conv$\{\mathbf{x} \in X$ and integer$\}$, and the second is that to find them it is necessary to solve an integer optimization problem. To overcome the first issue, DWI-models are solved in practice using column generation algorithms, which only pick attractive variables.

The solution of DWI-models with column generation algorithms is not a central topic in this monograph. Nevertheless, it is described succinctly as follows. The DWI-model has a *master problem*, defined by the first set of constraints in the DW-decomposition. The column generation algorithm starts with a *restricted master problem*, a model that has a restricted set of variables, which is optimized. The dual information from the restricted master problem is used in one or several *subproblems* to find the most attractive column to be inserted in the restricted master problem, which is then re-optimized. The iterative algorithm is repeated until no more attractive columns are found, yielding a solution that is provably optimal.

In each iteration, the optimal solution of the subproblem, which is stated in terms of the original variables $\mathbf{x}$, is an extreme point or an extreme ray of the set Conv$\{\mathbf{x} \in X$ and integer$\}$. Therefore, the subproblem is an integer optimization problem. This second issue may not be too hard to overcome. For instance, in the Cutting Stock

Problem, it amounts to finding integer solutions of a subproblem that is a knapsack problem.

Example 1.5 For the solution of the Gilmore and Gomory model (1.5)–(1.7), the restricted master problem is initialized with a set of columns (for instance, each column is a cutting pattern with multiple copies of the same item in quantities $\lfloor \frac{L}{\ell_i} \rfloor, \forall i$). Let $\bar{J}$ ($\bar{J} \subset J$) denote the set of cutting patterns in the restricted master problem, and $\bar{\mathbf{u}} = (\bar{u}_1, \bar{u}_2, \ldots, \bar{u}_m) \in \mathbb{R}_+^m$ be the corresponding optimal dual solution.

The subproblem is then used to find the most attractive column that does not belong to the restricted master problem (in $J \setminus \bar{J}$). The structure of a feasible column is well defined: $\mathbf{a}^j \in \mathbb{N}^m$ and it obeys (1.8). It will be attractive if its reduced cost, $\bar{c}_j = 1 - \bar{\mathbf{u}}^\top \mathbf{a}^j$, is negative. The most attractive cutting pattern is the column $\mathbf{a}^{min}$ with the most negative reduced cost $\bar{c}_{min} = 1 - \bar{\mathbf{u}}^\top \mathbf{a}^{min} = \min_{j \in J \setminus \bar{J}} (1 - \bar{\mathbf{u}}^\top \mathbf{a}^j)$. As all the columns in $\bar{J}$ have non-negative reduced costs, we may search over the entire set J. Furthermore, we may use the opposite function, to find the column j such that $\max_{j \in J} (\bar{\mathbf{u}}^\top \mathbf{a}^j - 1)$.

Therefore, given the dual optimal solution $\bar{\mathbf{u}}$ of a restricted master problem, the most attractive feasible column is provided by the solution of the following maximization knapsack subproblem:

$$\max \; z(\bar{\mathbf{u}}) := \bar{\mathbf{u}}^\top \mathbf{a}^j - 1 \tag{1.12}$$

$$\text{s. to} \; \mathbf{1}^\top \mathbf{a}^j \leq L \tag{1.13}$$

$$\mathbf{a}^j \in \mathbb{N}^m. \tag{1.14}$$

If the optimal solution of the knapsack subproblem is positive (corresponding to a negative reduced cost cutting pattern), the feasible column is added to the restricted master problem, which is re-optimized; otherwise, the current solution solves the linear relaxation of the Gilmore and Gomory model, because there are no more attractive columns. □

1.4 Duality and Bounds from Dual Feasible Solutions

Consider the following pair of primal and dual problems:

$$\begin{array}{ll} & \min \; z := \mathbf{c}^\top \mathbf{x} \\ (Primal) & \text{s. to} \; \mathbf{A}\mathbf{x} \geq \mathbf{b} \\ & \mathbf{x} \geq \mathbf{0}, \end{array} \qquad \begin{array}{ll} & \max \; z^D := \mathbf{b}^\top \mathbf{u} \\ (Dual) & \text{s. to} \; \mathbf{A}^\top \mathbf{u} \leq \mathbf{c} \\ & \mathbf{u} \geq \mathbf{0}, \end{array}$$

where $\mathbf{u} \in \mathbb{R}^m$ is the vector of dual decision variables.

The Weak Duality theorem states that, given a primal feasible solution $\hat{\mathbf{x}}$ and a dual feasible solution $\hat{\mathbf{u}}$, then $\mathbf{b}^\top \hat{\mathbf{u}} \leq z^D \leq z \leq \mathbf{c}^\top \hat{\mathbf{x}}$. It follows that any feasible solution $\hat{\mathbf{u}}$ to the dual problem provides a lower bound, $\mathbf{b}^\top \hat{\mathbf{u}}$, to the optimum solution of the primal problem, z. On the other hand, the Strong Duality theorem states that if the optima of the two problems are finite, then $z^D = z$.

Let us analyze what happens when we consider two primal models, one weak and one strong. For instance, recall that the primal minimization DWI-model that is stronger than the LP relaxation model yields a better optimum solution, i.e., $z_{LP} \leq z_{DWI} \leq z_{IP}$. According to the Strong Duality theorem, the optimal value of the dual maximization problem of the DWI-model will also be greater than or equal to the optimal value of the LP relaxation model. Therefore, one may expect to find feasible solutions of the dual maximization problem of the DWI-model that have an objective function value that is greater than or equal to z_{LP}.

Recall that we aim at finding feasible solutions of the dual models of strong DWI-models without enumerating any of the (exponentially many) columns that correspond to **all** the extreme solutions of the sets in the subproblem(s). Instead, by analyzing the structure of the dual of the DWI-models, we aim at deriving functions that provide dual feasible solutions that obey **all** the (exponentially many) constraints.

1.5 Examples

Two examples are presented below, illustrating that feasible solutions of the dual of a DWI-model can provide lower bounds that are better than trivial lower bounds. In Sect. 2.1, for the cutting stock problem, and in Sect. 4.3.3, for the vector packing problem, we will see how these dual feasible solutions are derived from DFF suited to each problem.

1.5.1 One-Dimensional Cutting Stock Problem

A trivial lower bound for the 1D-CSP results from calculating the minimum number of rolls of size L that are needed to place the sum of the sizes of all items:

$$LB_T = \lceil \sum_{i=1}^{m} b_i \ell_i / L \rceil.$$

The dual polytope of the Gilmore and Gomory model may have dual feasible solutions that provide better lower bounds. Recall that the dual polytope has all the (exponentially many) constraints that correspond to all feasible cutting patterns. As we will see, only *maximal patterns* (in which there is no room for any other item) are needed to define the dual polytope.

1.5 Examples

Example 1.6 A company has to deliver ten items with weight 0.4 and 40 items with weight 0.3. Each vehicle can carry a weight of 1. The lower bound $LB_T = \lceil (10 \times 0.4 + 40 \times 0.3)/1 \rceil = 16$, meaning that the sum of all weights is 16, and so, at least, 16 vehicles are needed. However, the optimal solution requires 17 vehicles. One optimal solution is having ten vehicles carrying items of sizes 0.4, 0.3 and 0.3, six vehicles carrying three items of size 0.3, and one vehicle carrying two items of size 0.3.

The set of all maximal feasible cutting patterns is $J_M = \{(a_1, a_2) \in \mathbb{N}^2 : 0.4 a_1 + 0.3 a_2 \leq 1\} = \{(2,0), (1,2), (0,3)\} \subset J$. The pair of primal-dual problems is as follows:

(Primal)
$$\begin{aligned} \min z &:= 1x_1 + 1x_2 + 1x_3 \\ \text{s. to} \quad & 2x_1 + 1x_2 \geq 10 \\ & 2x_2 + 3x_3 \geq 40 \\ & x_1, x_2, x_3 \geq 0, \end{aligned}$$

(Dual)
$$\begin{aligned} \max z^D &:= 10u_1 + 40u_2 \\ \text{s. to} \quad & 2u_1 \leq 1 \\ & 1u_1 + 2u_2 \leq 1 \\ & 3u_2 \leq 1 \\ & u_1, u_2 \geq 0. \end{aligned}$$

The space of the dual problem is depicted in Fig. 1.2. Note that cutting patterns that are not maximal would lead to redundant constraints.

The extreme points of the dual polytope are $O(0,0)$, $A(0, \frac{1}{3})$, $B(\frac{1}{3}, \frac{1}{3})$, $C(\frac{1}{2}, \frac{1}{4})$, $D(\frac{1}{2}, 0)$, respectively. Besides the extreme points, we consider the dual feasible solution $T(0.4, 0.3)$. The corresponding lower bounds are as follows:

$LB_O = \lceil (10 \times 0 + 40 \times 0)/1 \rceil = 0$ bins,
$LB_A = \lceil (10 \times 0 + 40 \times \frac{1}{3})/1 \rceil = 14$ bins,
$LB_B = \lceil (10 \times \frac{1}{3} + 40 \times \frac{1}{3})/1 \rceil = 17$ bins,
$LB_C = \lceil (10 \times \frac{1}{2} + 40 \times \frac{1}{4})/1 \rceil = 15$ bins,
$LB_D = \lceil (10 \times \frac{1}{2} + 40 \times 0)/1 \rceil = 5$ bins,
$LB_T = \lceil (10 \times 0.4 + 40 \times 0.3)/1 \rceil = 16$ bins. □

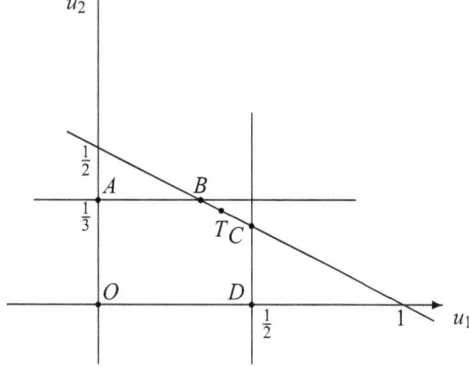

Fig. 1.2 Space of dual of Cutting Stock Problem Example

Note that the dual solution $\hat{u}_i = \ell_i/L, i = 1,\ldots,m$, is always feasible. It provides the lower bound $LB_T = \lceil \sum_{i=1}^{m} b_i \hat{u}_i \rceil$. In the example, it corresponds to the point $T(0.4, 0.3)$, which is not an extreme point of the dual polytope. There may even be cases, when all dual constraints have slack, where it is an interior point of the dual space. However, in other cases, the lower bound LB_T may be equal to the optimal solution of the 1D-CSP.

1.5.2 Vector Packing Problem

In the m-dimensional vector packing problem (mD-VPP), with $m \in \mathbb{N}\setminus\{0,1\}$, items with m independent dimensions (for instance, volume and weight in 2-dimensional problems) have to be packed into a minimum number of larger objects, which are m-dimensional bins. There are m capacity constraints, one for each dimension of the problem, i.e., the sum of the lengths of all packed items must not exceed the bin size in any of the m directions. The m-dimensional bins are all equal and have lengths L_d, $d = 1,\ldots,m$, and there are $n \in \mathbb{N}$ different items with lengths ℓ_{id} ($i = 1,\ldots,n$, $d = 1,\ldots,m$).

A trivial lower bound for the m-dimensional Vector Packing Problem, which amounts to applying the trivial lower bound for the 1D-CSP to all the dimensions and then taking the best value, is:

$$L_{VPP} = \max_{d=1,\ldots,m} \left\{ \left\lceil \sum_{i=1}^{n} \ell_{id}/L_d \right\rceil \right\}. \tag{1.15}$$

A DWI-model for the mD-VPP is similar to the Gilmore and Gomory model for the 1D-CSP (1.5)–(1.7), but the packing $\mathbf{a}^j \in \mathbb{N}^n$ is feasible if and only if

$$\sum_{i=1}^{n} d_i^j \times \ell_{id} \leq L_d, \quad d = 1,\ldots,m, \tag{1.16}$$

i.e., for each dimension, the sum of the lengths of the items packed in a bin does not exceed the length of the bin. Again, the dual polytope has dual feasible solutions that provide better lower bounds.

Example 1.7 A company has vehicles that can carry a volume of 4 and a weight of 5, and has to deliver four items with the following volumes and weights:

Item	1	2	3	4	Vehicle
Volume	2	3	1	2	4
Weight	3	2	4	1	5

1.5 Examples

Fig. 1.3 One optimal solution of the 2-dimensional VPP

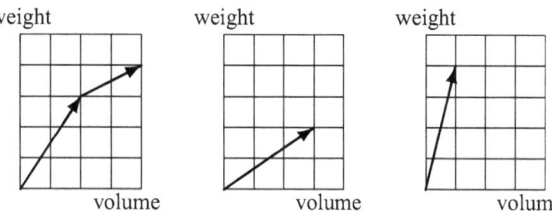

The trivial lower bound is $L_{VPP} = \max\{\lceil(2+3+1+2)/4\rceil, \lceil(3+2+4+1)/5\rceil\} = 2$; at least two vehicles are needed. However, the optimal solution of the 2-dimensional VPP instance requires three vehicles. One optimal solution is shown in Fig. 1.3. Vehicle 1 carries items 1 and 4, vehicle 2 carries item 2 and vehicle 3 carries item 3.

Let x_j be a decision variable that represents a feasible packing that assigns a set of items to a vehicle. The maximal packings for the 2D-VPP instance are:

		packings		
(v,w)	item	x_1	x_2	x_3
(2, 3)	1	1		
(3, 2)	2		1	
(1, 4)	3			1
(2, 1)	4	1		1
(4, 5)				

and the corresponding pair of primal-dual problems is as follows:

(Primal)
$$\begin{aligned}
\min z &:= x_1 + x_2 + x_3 \\
\text{s. to} \quad x_1 &\geq 1 \\
x_2 &\geq 1 \\
x_3 &\geq 1 \\
x_1 + x_3 &\geq 1 \\
x_1, x_2, x_3 &\geq 0,
\end{aligned}$$

(Dual)
$$\begin{aligned}
\max z^D &:= u_1 + u_2 + u_3 + u_4 \\
\text{s. to} \quad u_1 &+ u_4 \leq 1 \\
u_2 &\leq 1 \\
u_3 + u_4 &\leq 1 \\
u_1, u_2, u_3, u_4 &\geq 0.
\end{aligned}$$

Some feasible solutions of the dual space are $\hat{\mathbf{u}}_1 = (\frac{1}{2}, 1, 0, \frac{1}{2})^\top$, $\hat{\mathbf{u}}_2 = (1, 1, 0, 0)^\top$, $\hat{\mathbf{u}}_3 = (1, 1, 1, 0)^\top$ and $\hat{\mathbf{u}}_4 = (1, 0, 1, 0)^\top$. The dual solution $\hat{\mathbf{u}}_3$ is the one that provides the best lower bound, equal to 3, for this 2D-VPP instance. □

1.6 Related Literature

For a revision of integer and linear programming, as well as DW-decomposition, see the textbooks by Nemhauser and Wolsey (1998) or Bazaraa et al. (2010). DW-decomposition was first proposed by Dantzig and Wolfe (1960).

Geoffrion (1974) shows how to exploit problem structure in integer programming, and develops the general theory that relates to the bounds from the LP relaxation and from the models with subproblems without the integrality property. The theory is developed in the context of Lagrangian relaxation, which is closely related to DW-decomposition, as shown by Nemhauser and Wolsey (1998).

Gilmore and Gomory (1961) introduce the solution of the LP relaxation of the cutting stock problem using column generation, by solving an auxiliary integer knapsack problem. Valério de Carvalho (1999) introduces the arc-flow model for the cutting stock problem, and shows how the Gilmore and Gomory model can be derived as a non-negative combination of extreme rays, which are feasible solutions of the arc-flow model, and a null extreme point. Vance (1998) showed that the Gilmore and Gomory model can also be derived from a DW-decomposition of a model with a block angular structure. The derivation is left as an exercise.

Lueker (1983) uses the dual of the Gilmore and Gomory model to derive lower bounds for the bin-packing problem. In fact, the cutting stock problem and the bin-packing problem have structures that are very similar. The lower bound LB_T is due to Martello and Toth (1990) and the lower bound for the vector packing problem, LB_{VPP}, is due to Spieksma (1994).

1.7 Exercises

1. Consider the problem presented in Example 1.2, p. 5.

   (a) Apply a similar DW-decomposition, but considering the set X instead of the set Conv$\{\mathbf{x} : \mathbf{x} \in X \text{ and integer}\}$. Call Z_{DW} the value of the optimal solution of the DW-model.

   (b) Find the values of z_{IP}, z_{DWI}, z_{DW} and z_{LP}, and the relationship between them.

2. Consider the following LP problem with a block angular structure:

$$\begin{array}{rlllllll}
\max z := & 3\,x_1^1 & + & 5\,x_2^1 & + & 1\,x_1^2 & + & 2\,x_2^2 \\
\text{s. to} & 1\,x_1^1 & + & 2\,x_2^1 & + & 2\,x_1^2 & + & 1\,x_2^2 & \leq & 6 \\
& 3\,x_1^1 & + & 2\,x_2^1 & + & 1\,x_1^2 & + & 1\,x_2^2 & \leq & 8 \\
& 1\,x_1^1 & + & 2\,x_2^1 & & & & & \leq & 4 \\
& 1\,x_1^1 & & & & & & & \leq & 2 \\
& & & & & 3\,x_1^2 & + & 1\,x_2^2 & \leq & 3 \\
\end{array}$$

$$x_1^1,\ x_2^1,\ x_1^2,\ x_2^2 \geq 0.$$

1.7 Exercises

(a) Apply a DW-decomposition considering the following two bounded sets X^1 and X^2 in different subproblems:

$$X^1 = \{(x_1^1, x_2^1) \in \mathbb{R}^2 : x_1^1 + 2x_2^1 \leq 4,\ x_1^1 \leq 2,\ x_1^1, x_2^1 \geq 0\}$$
$$X^2 = \{(x_1^2, x_2^2) \in \mathbb{R}^2 : 3x_1^2 + x_2^2 \leq 3,\ x_1^2, x_2^2 \geq 0\},$$

(b) Draw the solution space of the two sets X^1 and X^2.
(c) Using an LP software package, solve the DW-model. Determine the optimal values of variables of the original LP model.

3. The parallel non-identical machine scheduling example, presented in Sect. 1.3.3, is an application of the generalized assignment problem (GAP). There is a set of jobs J, indexed by j, and a set of machines I, indexed by i, and we want to minimize the total processing cost of assigning jobs to machines, which is the sum of the costs of assigning each job to a machine, denoted by c_{ij}. The processing time of job j in machine i is denoted as p_{ij}, the time available in machine i is P_i.

The original model has binary decision variables x_{ij}, defined as follows:

$$x_{ij} = \begin{cases} 1, & \text{if job } j \text{ is processed in machine } i \\ 0, & \text{otherwise,} \end{cases}$$

and the IP model is as follows:

$$\min z_{IP} := \sum_{i \in I} \sum_{j \in J} c_{ij} x_{ij} \tag{1.17}$$

$$\text{s. to} \quad \sum_{j \in J} p_{ij} x_{ij} \leq P_i,\ \forall i \in I \tag{1.18}$$

$$\sum_{i \in I} x_{ij} = 1,\ \forall j \in J \tag{1.19}$$

$$x_{ij} \in \{0, 1\},\ \forall i \in I, j \in J. \tag{1.20}$$

There are two different DW-decompositions for the LP relaxation of this IP depending on the set of constraints that are placed in the subproblem, either constraints (1.18) or constraints (1.19).

(a) For each decomposition, describe the resulting master problem and subproblem.
(b) Discuss the quality of the bounds obtained from both DW-models, and compare them with the bound from the LP relaxation of the IP model above.
(c) Build both DW-models considering the data given in the following tables for a problem instance with two machines and four jobs. The times available in the machines are 10 and 12, respectively. The values of the processing

costs $C = [c_{ij}]$ and the processing times $P = [p_{ij}]$ are as follows:

$$C = \begin{array}{c|cccc} & 1 & 2 & 3 & 4 \\ \hline 1 & 15 & 12 & 10 & 9 \\ 2 & 8 & 4 & 8 & 6 \end{array} \qquad P = \begin{array}{c|cccc} & 1 & 2 & 3 & 4 \\ \hline 1 & 3 & 4 & 5 & 6 \\ 2 & 4 & 5 & 8 & 4 \end{array}$$

(d) Let $\mathbf{u} = (u_1, u_2, u_3, u_4)$ and $\mathbf{v} = (v_1, v_2)$ be the vectors of dual variables associated with the job and machine constraints, respectively. Check that $(\mathbf{u}, \mathbf{v}) = (15, 11, 15, 13, -5, -14)$ is a dual feasible solution for the strong DW-model. Which is the value of the corresponding lower bound?

(e) Check that the primal solution $y_2^1 = y_4^2 = 1$ and all the remaining $y_k^i = 0$ in Example 1.4 (p. 10) is feasible and optimal. Check also that this primal solution and the dual feasible solution obey the Complementary Slackness conditions.

4. Given bins of integer capacity L and a set of different item sizes $\ell_1, \ldots, \ell_m$, a feasible cutting pattern in a single roll can be modelled as a path in an acyclic directed graph with $L+1$ vertices. Consider a graph $G = (V, A)$ with $V = \{0, 1, 2, \ldots, L\}$ and $A = \{(i,j) : 0 \le i < j \le L \text{ and } \exists d \in \{1, \ldots, m\} \text{ with } j - i = \ell_d\}$, meaning that there is an arc between two vertices if there is an item of the corresponding size. There are additional arcs $(k, k+1)$, $k = 0, 1, \ldots, L-1$, corresponding to loss. A packing in a single bin is a path between vertices 0 and L.

The 1D-CSP is modelled as the problem of finding the minimum flow between vertex 0 and vertex L with additional constraints enforcing that the sum of the flows in the arcs of each order d must be greater than or equal to the corresponding demand b_d, $d = 1, \ldots, m$. The decision variables x_{ij}, associated with the arcs defined above, indicate the number of items of size $j - i$ placed in any roll at the distance of i units from the beginning of the roll. The variable x_{L0} can be seen as a feedback arc, from vertex L to vertex 0, and the model is as follows:

$$\min z := x_{L0} \tag{1.21}$$

$$\text{s. to } \sum_{(i,j) \in A} x_{ij} - \sum_{(j,k) \in A} x_{jk} = \begin{cases} -x_{L0}, & \text{if } j = 0 \\ 0, & \text{if } j = 1, \ldots, L-1 \\ x_{L0}, & \text{if } j = L \end{cases} \tag{1.22}$$

$$\sum_{(k,k+\ell_d) \in A} x_{k,k+\ell_d} \ge b_d, \quad d = 1, 2, \ldots, m \tag{1.23}$$

$$x_{ij} \ge 0, \; \forall (i,j) \in A \tag{1.24}$$

$$x_{ij} \text{ integer}, \; \forall (i,j) \in A. \tag{1.25}$$

(a) Apply a DW-decomposition to (1.21)–(1.24), keeping (1.22) and (1.24) in the subproblem and (1.23) in the master problem.
(b) Which are the extreme rays of the subproblem? Is there any extreme point?
(c) Consider a 1D-CSP instance with rolls of length 8 and items of lengths 4, 3 and 2, with order demands of 5, 4 and 8, respectively. The graph that

1.7 Exercises

represents the subproblem is shown in the figure (some arcs can be discarded and are not represented). Build the corresponding Gilmore and Gomory model.

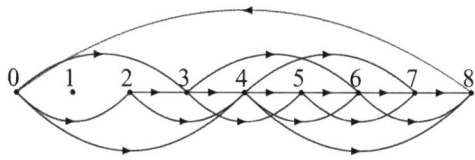

5. Given rolls of the same size L, each treated separately and indexed by k, $k \in K$, where K is a set of rolls that are sufficient to pack all the items, and clients with demands of d_i items of sizes ℓ_i, $0 < \ell_i \leq L$, $i \in I$, the cutting stock problem can be modelled using integer decision variables x_{ik}, which represent the number of times item i is cut from roll k, and binary decision variables y_k, with $y_k = 1$, if roll k is used, and 0, otherwise. The model is as follows:

$$\min z := \sum_{k \in K} y_k \qquad (1.26)$$

$$\text{s. to} \quad \sum_{k \in K} x_{ik} \geq d_i, \quad \forall i \in I, \qquad (1.27)$$

$$\sum_{i \in I} \ell_i x_{ik} \leq L y_k, \quad \forall k \in K, \qquad (1.28)$$

$$y_k \in \{0, 1\}, \quad \forall k \in K, \qquad (1.29)$$

$$x_{ik} \geq 0 \text{ and integer}, \quad \forall i \in I, \forall k \in K. \qquad (1.30)$$

The objective is to cut the minimum number of rolls to satisfy demand, and the first set of constraints enforces that the demand is satisfied, while the second imposes that the sum of the lengths of the items placed in one roll cannot exceed a function that takes the value of the length of the roll when the roll is used, and the value 0, otherwise.

(a) Apply a DW-decomposition to this model with a block angular structure, treating each roll as a separate entity, keeping (1.27) in the master problem, and each knapsack constraint of the set (1.28), together with (1.29)–(1.30), as a separate subproblem.
(b) Which is the meaning of the convexity constraint for each roll k in the reformulated model?
(c) Noting that all the rolls have equal size and their cutting stock patterns are identical, simplify the resulting DW-model, dropping the convexity constraints, to obtain the Gilmore and Gomory model.

Chapter 2
Classical Dual-Feasible Functions

2.1 Introduction

Dual-feasible functions (DFF) have been used to improve the resolution of different combinatorial optimization problems with knapsack inequalities, including cutting and packing, scheduling and network routing problems. They were used mainly to compute algorithmic lower bounds, but also to generate valid inequalities for integer programs. During a long time, the literature concerning these two applications of DFF was somehow *disconnected*. Functions defined for lower bounding were often referred to as *dual-feasible*, whereas the functions used to strengthen integer programming models were referred to as *superadditive and nondecreasing*. The relationship between these two families of functions is that the latter is a dominant family of DFF. Other designations are also used as for instance "*redundant function*" in the context of scheduling problems. These functions are in fact discrete DFF.

A dual-feasible function is defined formally as follows.

Definition 2.1 A function $f : [0, 1] \to [0, 1]$ is a *dual-feasible function*, if for any finite index set I of nonnegative real numbers $x_i \in \mathbb{R}_+$, $i \in I$, it holds that

$$\sum_{i \in I} x_i \leq 1 \implies \sum_{i \in I} f(x_i) \leq 1.$$

This implies immediately $f(0) = 0$ and $f(x) \leq 1/\lfloor 1/x \rfloor$ for all $x \in (0, 1]$.

Example 2.1 Figure 2.1 shows a parameter dependent staircase function $f : [0, 1] \to [0, 1]$, defined as

$$f(x) := \lfloor Cx \rfloor / \lfloor C \rfloor, \tag{2.1}$$

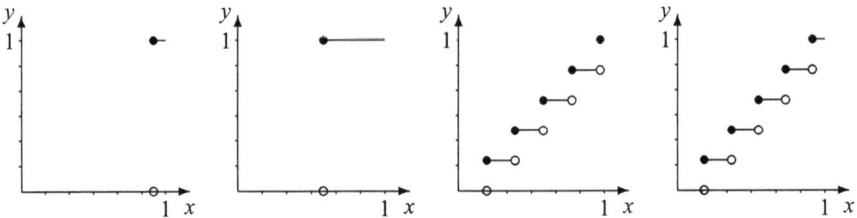

Fig. 2.1 Dual-feasible function (2.1) for parameter values $C \in \{\frac{12}{11}, \frac{12}{7}, \frac{36}{7}, \frac{60}{11}\}$

for several parameter values $C \geq 1$. Black filled circles mean that point belongs to the graph of the function, while white circles exclude that point. For instance, $f(\lfloor C \rfloor/C) = 1$, but $f(x) < 1$ for all $x \in [0, \lfloor C \rfloor/C)$. For the sake of conciseness, parameters are omitted wherever this is appropriate.

The function f is dual-feasible for any real parameter $C \geq 1$, as one can see as follows. One gets for any finite set I and numbers $x_i \in [0, 1]$ with $\sum_{i \in I} x_i \leq 1$ the estimation

$$s := \lfloor C \rfloor \times \sum_{i \in I} f(x_i) = \sum_{i \in I} \lfloor Cx_i \rfloor \leq \sum_{i \in I} Cx_i \leq C.$$

Since $\lfloor C \rfloor \times f(x_i) \in \mathbb{N}$, the sum is also integer, hence $s \leq \lfloor C \rfloor$.

Dual-feasible functions are generally defined in $[0, 1]$. However, using discrete values instead may lead to simpler formulations namely when the original input data is integer. This alternative formulation yields a so-called *discrete dual-feasible function* whose definition is given next.

Definition 2.2 A *discrete dual-feasible function* $f : \{0, 1, \ldots, d\} \to \{0, 1, \ldots, d'\}$ with $d, d' \in \mathbb{N} \setminus \{0\}$ is such that

$$\sum_{i \in I} x_i \leq d \implies \sum_{i \in I} f(x_i) \leq f(d) = d',$$

for any finite index set I of nonnegative integer numbers $x_i \in \mathbb{N}$, $i \in I$.

For every discrete DFF, there is an equivalent DFF defined in $[0, 1]$, and vice versa, as illustrated in the following example. Note, this does not mean a bijection.

Example 2.2 Let $g : [0, 1] \to [0, 1]$ be a DFF. Setting

$$f(x) := d' \times g(x/d),$$

for $x \in \{0, 1, \ldots, d\}$ yields a discrete DFF. Let for example

$$g(x) := \lceil \max\{0, kx - 1\} \rceil / (k - 1). \tag{2.2}$$

2.1 Introduction

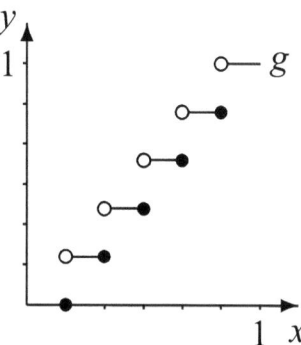

Fig. 2.2 The function g for $k = 6$

This function is a DFF for any $k \in \mathbb{N}$ with $k > 1$. Hence, $d' := k - 1$ yields the discrete DFF

$$f(x) := \left\lceil \max\left\{0, \frac{d'+1}{d}x - 1\right\} \right\rceil.$$

An illustration of the function g for $k = 6$ is provided in Fig. 2.2. □

Let $g : \{0, 1, \ldots, d\} \to \{0, 1, \ldots, d'\}$ be a discrete dual-feasible function with $d, d' \in \mathbb{N} \setminus \{0\}$ and $g(d) = d'$. A dual-feasible function $f : [0, 1] \to [0, 1]$ can be constructed as follows:

$$f(x) := g(\lfloor dx \rfloor)/d'. \tag{2.3}$$

The term *dual-feasible* comes from the alternative definition of these functions, which relies on the dual formulation of the well-known Gilmore and Gomory model for the cutting stock problem. In this context, a function f is said to be *dual-feasible* if it maps each x (the size of an item) to its corresponding value in a valid dual solution of the cutting stock problem. If f is a DFF, then assigning $u_i := f(\ell_i/L)$ for all $i \in \{1, \ldots, m\}$ yields a feasible solution for the dual problem (1.9)–(1.11), because (1.8) and the definition of a DFF ensure the validity of (1.10) and (1.11).

Example 2.3 Recall Example 1.6, p. 12, with items of sizes 0.4 and 0.3. Using the DFF g defined in (2.2), and shown in Fig. 2.3 for different values of the parameter k, the sizes of the items are mapped into the values indicated in the following table:

x	$g_2(x)$	$g_3(x)$	$g_4(x)$
0.4	0	$\frac{1}{2}$	$\frac{1}{3}$
0.3	0	0	$\frac{1}{3}$

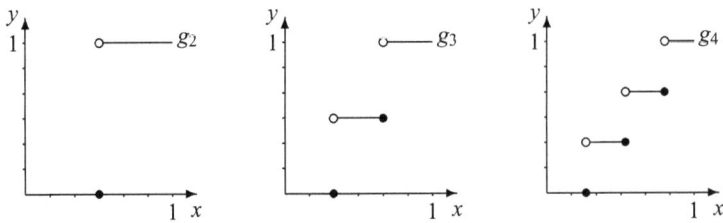

Fig. 2.3 DFF $g_k(x) := \lceil \max\{0, kx - 1\} \rceil / (k - 1)$, $k = 2, 3, 4$

Note that the DFF g_2, g_3 and g_4 provide the dual feasible solutions

$$O = (0,0), \ D = \left(\frac{1}{2}, 0\right) \text{ and } B = \left(\frac{1}{3}, \frac{1}{3}\right),$$

respectively, shown in Example 1.6.

Each DFF in the family provides a lower bound, whose computation is not expensive at all. For this instance, the general expression for the values of the lower bounds is

$$LB_{g_k} := \lceil (10 \times g_k(0.4) + 40 \times g_k(0.3))/1 \rceil,$$

for $k > 1$. □

Hence, dual-feasible functions lead to valid lower bounds for the cutting stock problem, and there is always one such function whose corresponding bound is as strong as the continuous bound achieved with the Gilmore and Gomory model (1.5)–(1.7) presented in Sect. 1.3.1, because there is a DFF f, such that $\mathbf{b}^\top \mathbf{u} = \sum_{i=1}^{m} b_i \times f(\ell_i/L)$ equals the optimal objective function value of the continuous relaxation of (1.5)–(1.7). To see this, assume without loss of generality that $\ell_i \neq \ell_j$ for all $i \neq j$, and let $\hat{\mathbf{u}}$ be an optimal solution of (1.9)–(1.11). A suitable DFF f is obtained by setting $f(x) := \hat{u}_i$ for $x = \ell_i/L$ ($i = 1, \ldots, m$) and $f(x) := 0$ for all the remaining points, i.e., $f(x) := 0$ for all $x \in [0, 1] \setminus \{\ell_i/L | i \in \{1, \ldots, m\}\}$.

2.2 Properties

2.2.1 Maximality

Despite the large number of dual-feasible functions that may be defined, only those that are non-dominated are interesting since they yield the best lower bounds and

2.2 Properties

strongest valid inequalities. To be non-dominated, a dual-feasible function must be superadditive and nondecreasing.

Definition 2.3 A function f is *superadditive* if for all x, y with $x, y, x + y$ belonging to the domain of f, it holds that $f(x + y) \geq f(x) + f(y)$.

Proposition 2.1 *If a function $f : [0, 1] \to [0, 1]$ is superadditive, then f is a DFF.*

Proof The range and the superadditivity imply monotonicity and

$$\sum_{i \in I} f(x_i) \leq f\left(\sum_{i \in I} x_i\right) \leq f(1) \leq 1$$

for any finite index set I of nonnegative real numbers x_i with $\sum_{i \in I} x_i \leq 1$. □

Note that a superadditive function may be dominated by another superadditive function. A non-dominated dual-feasible function is said to be *maximal* as defined next.

Definition 2.4 A DFF $f : [0, 1] \to [0, 1]$ is a *maximal dual-feasible function* (MDFF), if there is no other DFF $g : [0, 1] \to [0, 1]$ with $g(x) \geq f(x)$ for all $x \in [0, 1]$. Similarly, a discrete DFF $h : \{0, 1, \ldots, d\} \to \{0, 1, \ldots, d'\}$ is *maximal*, if there is no other discrete DFF $k : \{0, 1, \ldots, d\} \to \{0, 1, \ldots, d'\}$ with $k(x) \geq h(x)$ for all $x \in \{0, 1, \ldots, d\}$.

Example 2.4 The following function $f_{BJ,1}(x; C)$ is a MDFF for all parameter values $C \geq 1$. It is defined as

$$f_{BJ,1}(x; C) := \left(\lfloor Cx \rfloor + \max\left\{0, \frac{\mathsf{frac}(Cx) - \mathsf{frac}(C)}{1 - \mathsf{frac}(C)}\right\}\right) / \lfloor C \rfloor, \qquad (2.4)$$

where $\mathsf{frac}(\cdot)$ denotes the non-integer part of its argument, i.e. $\mathsf{frac}(C) \equiv C - \lfloor C \rfloor$. Function $f_{BJ,1}$ is continuous and piecewise linear for all C, as illustrated in Fig. 2.4. If $C \in \mathbb{N}$, then $f_{BJ,1}$ becomes the identity function. Otherwise, there are $\lceil C \rceil$ many intervals without a slope, and the slope in the other intervals increases with $\mathsf{frac}(C)$.

Note that, without the max-expression, the non-maximal staircase DFF $f : [0, 1] \to [0, 1]$, defined in formula (2.1), would be obtained. In the open intervals

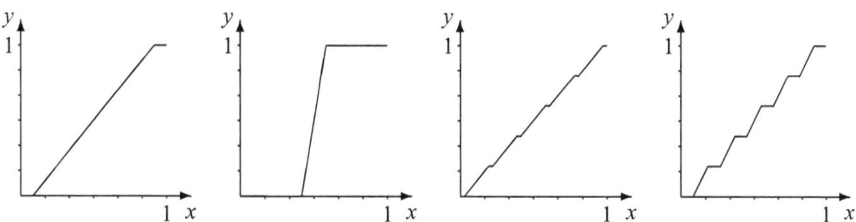

Fig. 2.4 Maximal dual-feasible function $f_{BJ,1}$ for parameter values $C \in \{\frac{12}{11}, \frac{12}{7}, \frac{36}{7}, \frac{60}{11}\}$

where $f_{BJ,1}$ is strictly monotone, one has $f(x) < f_{BJ,1}(x)$. Only outside these intervals $f(x) = f_{BJ,1}(x)$ holds. □

Several properties characterize the MDFF. They have to be nondecreasing, superadditive, and also *symmetric*, as stated formally in the following theorem.

Theorem 2.1 *A function $f : [0, 1] \to [0, 1]$ is a MDFF if and only if the following conditions hold:*

1. *f is monotonely increasing, i.e. $f(x) \leq f(y)$ if $x \leq y$;*
2. *f is superadditive;*
3. *f is symmetric in the sense $f(x) + f(1 - x) = 1$, $\forall x \in [0, 1]$;*
4. *$f(0) = 0$.*

Some of these conditions imply others. Therefore, to prove that a given function is a MDFF, much weaker prerequisites are in fact sufficient.

Theorem 2.2 *A function $f : [0, 1] \to \mathbb{R}_+$ fulfilling the following conditions is a MDFF:*

$$f(0) = 0; \tag{2.5}$$

$$f(x) + f(1 - x) = 1, \ \forall x \in [0, 1/2]; \tag{2.6}$$

$$f(x_1 + x_2) \geq f(x_1) + f(x_2), \ \forall x_1, x_2 \text{ with}$$
$$0 < x_1 \leq x_2 < 1/2 \text{ and } x_1 + x_2 \leq 2/3. \tag{2.7}$$

Note that to prove that a real function f is a MDFF using this theorem, it is not necessary to prove that $f(x) \leq 1$ for all $x \in [0, 1]$. Indeed, this follows from $f(x) \geq 0$ and the symmetry condition (2.6). Therefore, the range of f is only required to be part of $\mathbb{R}_+$.

When verifying whether a given function f is a MDFF, the main difficulty in the application of Theorems 2.1 and 2.2 is usually related to the test of its superadditivity. An approach for this test consists in resorting to the following function $g : (0, 1/2)^2 \to \mathbb{R}$:

$$g(x_1, x_2) := f(x_1 + x_2) - f(x_1) - f(x_2). \tag{2.8}$$

The function f obeys the superadditivity condition (2.7) if and only if $g(x_1, x_2) \geq 0$ for all x_1, x_2 according to (2.7). To check this, the extreme points of g can be sought. If g is differentiable, then only the critical points, i.e. those with $\nabla g(x_1, x_2) = \mathbf{0}$, may be extreme points. If f is differentiable only in some smaller intervals inside $(0, 1)$, then analyzing the function (2.8) requires also to check separately points where $f'(x_1), f'(x_2)$ or $f'(x_1 + x_2)$ does not exist.

Lemma 2.1 *If the function $f : [0, 1] \to [0, 1]$ fulfils the symmetry condition (2.6), and if it is differentiable in the interval $(0,1)$, then the point $\left(\frac{1}{3}, \frac{1}{3}\right)$ is a critical point for $g(x_1, x_2)$ in (2.8).*

2.2 Properties

According to Lemma 2.1, the superadditivity of f should always be checked for $x_1 = x_2 = 1/3$ first. If f is symmetric, then $2 \times f(1/3) \leq f(2/3)$ holds if and only if $f(1/3) \leq 1/3$. A drawback of this approach is that the function $g(x_1, x_2)$ in (2.8) may be complicated and have infinitely many critical points. Since every critical point is potentially an extreme point, checking the superadditivity of the function f by testing the nonnegativity of g at its extreme points relies on exploring all the critical points of g.

2.2.2 Maximality of Convex Functions

The particular case of convex functions is addressed next.

Definition 2.5 Let $D \neq \emptyset$ be a convex set. A function $f : D \to \mathbb{R}$ is *convex*, if for all $x_1, x_2 \in D$ and $\lambda \in (0, 1)$, it holds that

$$f(\lambda x_1 + (1 - \lambda)x_2) \leq \lambda \times f(x_1) + (1 - \lambda) \times f(x_2).$$

The superadditivity of these functions (not necessarily bounded to domain and range $[0, 1]$) is established through the following lemma.

Lemma 2.2 *Let $b > 0$ be a constant. If a function $f : [0, b] \to \mathbb{R}$ is convex on $[0, b]$ and if $f(0) \leq 0$, then f is superadditive.*

As a corollary, we obtain the following result that allows for the simple identification of many MDFF.

Lemma 2.3 *If $f : [0, 1] \to [0, 1]$ with $f(0) = 0$ fulfills the symmetry condition (2.6), and if it is convex on $[0, 1/2]$, then f is a MDFF.*

Using Lemma 2.3, it is easy to see for $C \leq 2$ that the function $f_{BJ,1}(x; C)$ in (2.4) is a MDFF.

2.2.3 Extremality

To get quickly the strongest bounds and inequalities from dual-feasible functions, maximality is not enough. Consider for example the case of the 1D-CSP as defined in Sect. 1.3.1. If we are given three MDFF $f, g, h : [0, 1] \to [0, 1]$, such that $2f(x) = g(x) + h(x)$ for all $x \in [0, 1]$ (with $x_i = l_i/L$, $i = 1, \ldots, m$), then the lower bound obtained with these dual feasible functions equals

$$\left(\sum_{i=1}^{m} b_i \times g(x_i) + \sum_{i=1}^{m} b_i \times h(x_i) \right)/2,$$

and hence

$$\sum_{i=1}^{m} b_i \times f(x_i) \leq \max\left\{\sum_{i=1}^{m} b_i \times g(x_i), \sum_{i=1}^{m} b_i \times h(x_i)\right\}.$$

In that case, either g or h leads to a bound, which is not worse than the one obtained by f. Therefore, in order to avoid unnecessary calculations, the use of f is superfluous, and it is in practice important, apart from maximality, to know whether a dual-feasible function leads to solutions, or not, which are always dominated or achieved by another DFF.

Given a convex set $S \neq \emptyset$, a point $e \in S$ is an extreme point if $2e = s_1 + s_2$ with $s_1, s_2 \in S$ implies $e = s_1 = s_2$. A similar definition can be adopted for dual-feasible functions.

Definition 2.6 A MDFF f is an *extreme maximal dual-feasible function (EMDFF)*, if for any MDFF g, h with

$$2 \times f(x) = g(x) + h(x), \quad \forall x \in [0, 1],$$

it follows that $f \equiv g$.

For any non-extreme MDFF f, and any finite set I and $x_i \in [0, 1]$, with $i \in I$, there is another MDFF g such that

$$\sum_{i \in I} g(x_i) \geq \sum_{i \in I} f(x_i).$$

If f is a non-extreme MDFF, then there are other MDFF g, h with

$$\sum_{i \in I} f(x_i) = \frac{1}{2} \times \left(\sum_{i \in I} g(x_i) + \sum_{i \in I} h(x_i)\right).$$

and either both summands equal $\sum_{i \in I} f(x_i)$ or one is larger. Clearly, it makes no sense to analyze the case of non-maximal DFF, since they are dominated by at least one MDFF by definition.

Example 2.5 Consider the following MDFF $f_{CCM,1}(x; C)$ illustrated in Fig. 2.5:

$$f_{CCM,1}(x; C) := \begin{cases} \lfloor Cx \rfloor / \lfloor C \rfloor, & \text{if } 0 \leq x < 1/2, \\ 1/2, & \text{if } x = 1/2, \\ 1 - f_{CCM,1}(1-x; C), & \text{if } 1/2 < x \leq 1. \end{cases} \quad (2.9)$$

This function is extreme for all the feasible values of its parameter. As an example, we provide the proof for $1 \leq C < 3$. Let $g, h : [0, 1] \to [0, 1]$ be MDFF

2.2 Properties

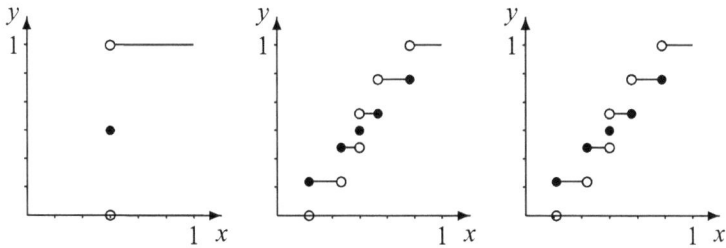

Fig. 2.5 MDFF $f_{CCM,1}$ for parameter values $1 \leq C \leq 2$, $C = \frac{36}{7}$ and $C = \frac{60}{11}$

with $2f_{CCM,1} \equiv g + h$. The conditions for a DFF to be maximal imply

$$g(0) = h(0) = 0 \text{ and } g(1/2) = h(1/2) = 1/2.$$

Because of the symmetry condition (2.6), we only need to show that $g(x) = h(x)$ for all $x \in (0, 1/2)$:

- if $Cx < 1$, then $f_{CCM,1}(x) = 0$, and hence $g(x) = h(x) = 0$ due to the range of g and h;
- if $2 < C < 3$ and $1/C \leq x < 1/2$, then $f_{CCM,1}(x) = 1/2$. Since $g(x), h(x) \leq 1/2$ for $x \leq 1/2$, it follows that $g(x) = h(x) = 1/2$, for $1/C \leq x < 1/2$. □

As discussed above, a non-extreme MDFF should not be used to obtain bounds, because that would yield only a convex combination. However, even an extreme MDFF needs not to yield corners of the dual polyhedron.

Example 2.6 The identity function is an EMDFF. Consider the one-dimensional cutting stock instance that $b \in \mathbb{N}$ items of length 5 have to be cut from initial material of length 9. The identity function yields the bound $\frac{5}{9}b$ according to the dual variable $u := 5/9$. However, any $u \in [0, 1]$ would have been a feasible dual variable, and $5/9$ is in the inner of this region. □

2.2.4 Extremality of Convex Functions

Proving or disproving the extremality of a DFF is usually non-trivial. When the function is convex, we may use the following results to make these proofs easier.

Theorem 2.3 *Let $f : [0, 1] \to [0, 1]$ be a MDFF such that f is convex on $[0, 1/2]$. For given values a, b, c with $0 < a < b < 1/2$ and $c > 0$, if f has a continuous second derivative on $[a, b]$ and $f''(x) \geq c$ for all $x \in [a, b]$, then f is not extreme.*

Proof The proof consists of constructing two different MDFF $g, h : [0, 1] \to [0, 1]$ with $2f \equiv g + h$. These functions g, h are built by perturbing f inside (a, b) such that g, h remain convex on $[0, 1/2]$. We will have to choose an enough small $\lambda > 0$. Define the functions $l : \mathbb{R} \to [0, 2]$ and $g, h : [0, 1] \to [0, 1]$ according to

$$l(x) := \begin{cases} 1 - \cos(2\pi x), & \text{if } 0 < x < 1, \\ 0, & \text{otherwise}; \end{cases}$$

$$g(x) := \begin{cases} f(x) + \lambda \times l(\frac{x-a}{b-a}), & \text{if } 0 \le x \le 1/2, \\ 1 - g(1-x), & \text{otherwise}; \end{cases}$$

$$h(x) := 2f(x) - g(x).$$

The function l is once continuously differentiable. It follows that $f(x) = g(x) = h(x)$ for all $x \in [0, 1] \setminus (a, b)$, all these functions are twice continuously differentiable in (a, b) and once in an environment $U \subseteq (0, 1/2)$ of the closed interval $[a, b]$. Since g and h are symmetric, it remains to show, according to Lemma 2.3, that g', h' are monotonely increasing on $[a, b]$ and $f \neq g$. One obtains inside the interval (a, b) the following derivatives:

$$g'(x) = f'(x) + \lambda \times \frac{2\pi}{b-a} \sin\left(\frac{2\pi}{b-a}(x-a)\right);$$

$$g''(x) = f''(x) + \lambda \times \left(\frac{2\pi}{b-a}\right)^2 \cos\left(\frac{2\pi}{b-a}(x-a)\right);$$

$$h''(x) = 2f''(x) - g''(x)$$

If $a < x < \frac{a+b}{2}$ then $g'(x) > f'(x)$. Therefore, $f \neq g$.

To show the convexity of g in $[0, 1/2]$, one may use the estimation $g''(x) \ge c - (\frac{2\pi}{b-a})^2 \lambda \ge 0$ for $\lambda \le (\frac{b-a}{2\pi})^2 c$ inside the interval (a, b). Due to $g''(x) \ge 0$ in (a, b) the derivative g' is monotonely rising in (a, b). Because of $g'(a) = f'(a)$ and $g'(b) = f'(b)$ and the continuous differentiability of g in U the derivative g' is also monotone in U. Therefore, g is convex in the needed interval $[0, 1/2]$.

Regarding h, the same considerations are valid. □

Because of Theorem 2.3, many functions $f : [0, 1] \to [0, 1]$, which are convex on $[0, 1/2]$, can be proved to be non-extreme. In practice, while most of the MDFF that may be built through Lemma 2.3 will be non-extreme, with Theorem 2.3, we can avoid the search for the "best" MDFF in a large part of the search area. Because of Theorem 2.3, to build an EMDFF by Lemma 2.3, one must ensure that f is linear in all intervals where f'' exists.

The following lemma may simplify the proofs that a certain MDFF is extreme in the case where the function is piecewise linear. This restriction comes from part (a) of Lemma 2.4.

Lemma 2.4 *Let a, b be given values with $0 \le a < b \le 1/2$, and a MDFF $f : [0, 1] \to [0, 1]$ such that, for all $x, y \in [a, b]$, it holds that $f(x + y) = f(x) + f(y)$. We have:*

2.2 Properties

(a) $f(x) = \frac{f(b)-f(a)}{b-a} \times (x-a) + f(a)$, for all $x \in [a,b]$, i.e., f is linear in that interval;
(b) if $g, h : [0, 1] \to [0, 1]$ are MDFF with

$$2f(x) = g(x) + h(x), \forall x \in [0, 1],$$

and $g(a) = f(a)$ and $g(b) = f(b)$, then

$$g(x) = f(x), \forall x \in [a, b].$$

Proof

(a) Let $a_0 := a$ and $b_0 := b$. For arbitrary $\zeta \in [a, b]$ it will be shown that $f(\zeta) = \frac{f(b)-f(a)}{b-a} \times (\zeta - a) + f(a)$. Clearly, this holds for $\zeta \in \{a, b\}$. Assume that for some $x, y \in [a, b]$ a similar equation holds. Since $f(x+y) = 2 \times f(\frac{x+y}{2})$ due to the prerequisites, it follows that

$$f\left(\frac{x+y}{2}\right) = \frac{f(x)+f(y)}{2}$$

$$= \frac{f(b)-f(a)}{2(b-a)} \times (x - a + y - a) + f(a)$$

$$= \frac{f(b)-f(a)}{b-a} \times \left(\frac{x+y}{2} - a\right) + f(a).$$

Hence, the proposition is also true for $(x+y)/2$. To prove it for the given ζ, an interval interlock is used. For $n = 0, 1, 2, \ldots$, it is constructed as follows: if $2\zeta > a_n + b_n$ then let $a_{n+1} := (a_n + b_n)/2$ and $b_{n+1} := b_n$, otherwise $a_{n+1} := a_n$ and $b_{n+1} := (a_n + b_n)/2$. That yields

$$a_0 \leq \cdots \leq a_n \leq \cdots \leq \zeta \leq \cdots \leq b_n \leq \cdots \leq b_0$$

and $\lim_{n \to \infty} (b_n - a_n) = 0$, where the desired equation holds for all a_n and b_n. Every maximal dual-feasible function is monotonely increasing. Therefore,

$$\frac{f(b)-f(a)}{b-a} \times (a_n - a) \leq f(\zeta) - f(a)$$

$$\leq \frac{f(b)-f(a)}{b-a} \times (b_n - a)$$

for all $n \in \mathbb{N}$. The difference between the right and left part of the inequality tends to zero. Therefore, f is continuous at ζ, and the proposition (a) is true.

(b) If for certain numbers $x, y \in [a, b]$, the equations $g(x) = f(x)$ and $g(y) = f(y)$ are valid, then the superadditivity condition yields

$$g(x+y) \geq f(x+y) \quad \text{and} \quad h(x+y) \geq f(x+y),$$

and consequently, $g(x + y) = f(x + y) = 2 \times f(\frac{x+y}{2})$. Therefore, $g\left(\frac{x+y}{2}\right) \leq f\left(\frac{x+y}{2}\right)$ and $h\left(\frac{x+y}{2}\right) \leq f\left(\frac{x+y}{2}\right)$. That implies $g\left(\frac{x+y}{2}\right) = f\left(\frac{x+y}{2}\right)$. Choose any $\zeta \in [a, b]$. The proposition $g(\zeta) = f(\zeta)$ can be shown analogously to part (a). The monotone sequences (a_n) and (b_n) are defined as above. Since $g(a_n) = f(a_n)$ and $g(b_n) = f(b_n)$ for all $n \in \mathbb{N}$, it follows that $g(\zeta) = f(\zeta)$, because g, being a MDFF, is monotone, and f is continuous in $[a, b]$. □

The following example shows in a simplified way how part (b) of Lemma 2.4 can be used to prove that a given maximal dual-feasible function is extreme. Moreover, the example also demonstrates the difficulty in the analysis of a parameter dependent maximal dual-feasible function that is extreme for some parameter values and non-extreme for others.

Example 2.7 The function $f_{BJ,1}(x; C)$ defined in (2.4) is extreme for $C \in \mathbb{N}$ and for $C \geq 2$, but not for $1 < C < 2$. Only part of the proof is provided. The remaining part is left as an exercise.

For $C \in \mathbb{N}$, the assertion follows almost immediately from Lemma 2.4, because $f_{BJ,1}$ becomes the identity function. Suppose, $g, h : [0, 1] \to [0, 1]$ are MDFF with $2 \times f_{BJ,1} \equiv g + h$. Therefore, we have $g(0) = h(0) = 0$ and $g(1/2) = h(1/2) = 1/2$. Setting $a := 0$ and $b := 1/2$ in Lemma 2.4(b) closes the proof.

For $1 < C < 2$, the function $f_{BJ,1}(x; C)$ is a convex combination of $f_{BJ,1}(x; \tilde{C})$ and another continuous MDFF $g : [0, 1] \to [0, 1]$, where $\tilde{C} = 2C$ if $1 < C < 4/3$, and $\tilde{C} = \frac{4C}{C+2}$ if $4/3 \leq C < 2$. If $1 < C < 4/3$, then

$$g(x) = \begin{cases} 0, & \text{if } 0 \leq x \leq 1 - \frac{1}{C}, \\ \frac{(Cx+1-C)(4-3C)}{(3-2C)(2-C)}, & \text{if } 1 - \frac{1}{C} \leq x \leq \frac{1}{2C}, \\ \frac{4Cx+2-3C}{4-2C}, & \text{if } \frac{1}{2C} \leq x \leq 1 - \frac{1}{2C}, \\ 1 - g(1-x), & \text{if } 1 - \frac{1}{2C} \leq x \leq 1, \end{cases}$$

and if $4/3 \leq C < 2$ then

$$g(x) = \begin{cases} 0, & \text{if } 0 \leq x \leq 1 - \frac{1}{C}, \\ \frac{2Cx+2-2C}{2-C}, & \text{if } 1 - \frac{1}{C} \leq x \leq \frac{3C-2}{4C}, \\ \frac{1}{2}, & \text{if } \frac{3C-2}{4C} \leq x \leq \frac{C+2}{4C}, \\ 1 - g(1-x), & \text{if } \frac{C+2}{4C} \leq x \leq 1. \end{cases}$$

□

2.3 Generating One-Dimensional Dual-Feasible Functions

In this section, we show how to build non-trivial dual-feasible functions from simple superadditive functions. We address first the simple case of linear combination, and then, we explore the properties of composed dual-feasible functions. We explain how to define a maximal dual-feasible function from a non-maximal function, while alternative approaches are explored at the end.

2.3.1 Linear Combination

The simplest way to generate a dual-feasible function is to combine linearly two functions. First, superadditivity is preserved: if f and g are superadditive functions, then $\alpha f + \beta g$ (with $\alpha, \beta \in \mathbb{R}_+$) will be superadditive too. Similarly, combining linearly two maximal dual-feasible functions does not affect maximality as stated next.

Proposition 2.2 *If f and g are discrete MDFF with the same domain $\{0, 1, \ldots, d\}$, $d \in \mathbb{N} \setminus \{0\}$, then $h := \alpha f + \beta g$ ($\alpha, \beta \in \mathbb{N}, \alpha + \beta > 0$) is a discrete MDFF too.*

Proof Since $\alpha, \beta \geq 0$, the range of f and g implies $h(x) \geq 0$ for all $x \in \{0, 1, \ldots, d\}$. We show that h is symmetric and superadditive. Since f and g are discrete MDFF, one has for any $x \in \{0, 1, \ldots, d\}$ the following:

$$0 < f(d) = f(x) + f(d-x) \quad \text{and} \quad 0 < g(d) = g(x) + g(d-x)$$

Therefore,

$$0 < \alpha \times f(d) + \beta \times g(d) = h(d)$$
$$= \alpha \times f(x) + \alpha \times f(d-x) + \beta \times g(x) + \beta \times g(d-x)$$
$$= h(x) + h(d-x),$$

and hence h is symmetric. To verify the superadditivity, choose any $x, y \in \{0, 1, \ldots, d\}$ with $x + y \leq d$. Since f and g are superadditive, we have

$$f(x+y) \geq f(x) + f(y) \quad \text{and} \quad g(x+y) \geq g(x) + g(y).$$

That implies

$$h(x+y) = \alpha \times f(x+y) + \beta \times g(x+y)$$
$$\geq \alpha \times f(x) + \alpha \times f(y) + \beta \times g(x) + \beta \times g(y)$$
$$= h(x) + h(y).$$

□

Note that, when functions with domain and range [0, 1] are considered, αf makes little sense, and $\alpha f + \beta g$ (with $\alpha, \beta \in \mathbb{R}_+$) should be replaced by a convex combination of f and g. In practice, Proposition 2.2 is meaningful mainly for discrete functions.

Preserving superadditivity does not imply that maximality is preserved too. For instance, while $\lfloor f \rfloor$ and $\min\{f, g\}$ remain superadditive, function $\min\{f, g\}$ is not maximal, unless $f = g$, nor is $\lfloor f \rfloor$ ($x \mapsto \lfloor x \rfloor$ is not even a MDFF).

Extremality was only introduced for maximal dual-feasible functions with domain and range [0, 1]. For these functions, and by definition, if one has two different maximal dual-feasible functions $f, g : [0, 1] \to [0, 1]$, then $h := \frac{\alpha}{\alpha+\beta}f + \frac{\beta}{\alpha+\beta}g$ with $\alpha, \beta > 0$ cannot yield an extreme maximal dual-feasible function. Only the trivial combination with $\beta = 0$ and f being extreme, or $\alpha = 0$ and g being extreme, yields an extreme maximal dual-feasible function.

2.3.2 Composition

Composing functions is another possible approach to build dual-feasible functions. As with linear combinations, superadditivity and maximality also remain with composition.

Proposition 2.3 *Let $f, g : [0, 1] \to [0, 1]$ be two MDFF. The composed function $f(g(x))$ is also a MDFF.*

Proof Let $x, y \in [0, 1]$ with $x + y \leq 1$. We have

$$f(g(0)) = f(0) = 0,$$
$$f(g(1-x)) = f(1 - g(x)) = 1 - f(g(x)), \text{ and}$$
$$g(x + y) \geq g(x) + g(y),$$

and hence

$$f(g(x+y)) \geq f\big(g(x+y) - g(x) - g(y)\big) + f(g(x)) + f(g(y))$$
$$\geq f(g(x)) + f(g(y)).$$

The composition $f(g(x))$ fulfills the sufficient conditions to be a MDFF. Furthermore, since $f(g(x))$ is superadditive and $f(g(x)) \geq 0$, this function is also non-decreasing. □

On the contrary, composing maximal dual-feasible functions that are extreme does not always yield a function that is extreme as illustrated next.

Proposition 2.4 *The composition of extreme MDFF is not necessarily extreme.*

2.3 Generating One-Dimensional Dual-Feasible Functions

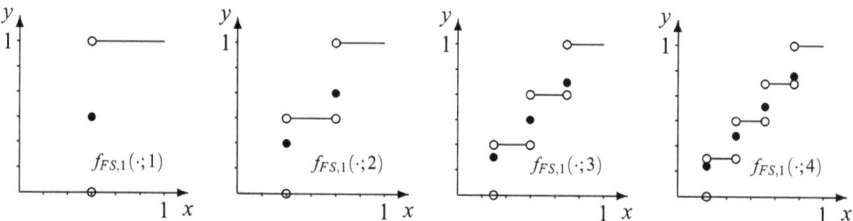

Fig. 2.6 MDFF $f_{FS,1}$ for parameter values $k \in \{1, \ldots, 4\}$

Proof Consider the MDFF $f_{FS,1}(x;k) : [0,1] \to [0,1]$

$$f_{FS,1}(x;k) := \begin{cases} x, & \text{if } (k+1) \times x \in \mathbb{N}, \\ \lfloor (k+1) \times x \rfloor / k, & \text{otherwise}, \end{cases} \quad (2.10)$$

with $k = 2$, and $f_{BJ,1}(x;C)$ defined in (2.4) with $C = 9/2$, and let $f := f_{FS,1}(f_{BJ,1}(\cdot))$. Both $f_{FS,1}$ and $f_{BJ,1}$ are extreme. One has $f_{BJ,1}(10/27) = 1/3$, and $f_{BJ,1}(x) \neq 1/3$ for $x \in [0,1] \setminus \{10/27\}$.

The function $f_{FS,1}$ is illustrated in Fig. 2.6 for several parameter values.

Because of $f_{FS,1}(x) = 0$ for $0 \leq x < 1/3$, $f_{FS,1}(1/3) = 1/3$ and $f_{FS,1}(x) = 1/2$ for $1/3 < x < 2/3$, it follows that

$$f(x) = \begin{cases} 0, & \text{if } 0 \leq x < 10/27, \\ 1/3, & \text{if } x = 10/27, \\ 1/2, & \text{if } 10/27 < x < 17/27, \\ 1 - f(1-x), & \text{if } 17/27 \leq x \leq 1. \end{cases}$$

Let $g, h : [0,1] \to [0,1]$ be two MDFF defined as

$$g(x) := \begin{cases} 0, & \text{if } 0 \leq x \leq 10/27, \\ 1/2, & \text{if } 10/27 < x < 17/27, \\ 1, & \text{if } 17/27 \leq x \leq 1, \end{cases} \quad \text{and } h(x) := \begin{cases} 0, & \text{if } 0 \leq x < 10/27, \\ 1/2, & \text{if } 10/27 \leq x \leq 17/27, \\ 1, & \text{if } 17/27 < x \leq 1. \end{cases}$$

Then, $f(x) = \frac{1}{3}g(x) + \frac{2}{3}h(x)$, for all $x \in [0,1]$, and hence f cannot be extreme. □

2.3.3 Symmetry

Maximal dual-feasible functions can be built from superadditive functions by keeping the images of the values smaller than $1/2$, and by computing the images of the values larger than $1/2$ by symmetry.

Theorem 2.4 *Let* $f : [0, 1] \to [0, 1]$ *be a superadditive function. The following function* $\hat{f} : [0, 1] \to [0, 1]$, *defined as*

$$\hat{f}(x) := \begin{cases} f(x), & \text{if } 0 \leq x < 1/2, \\ 1/2, & \text{if } x = 1/2, \\ 1 - f(1-x), & \text{if } 1/2 < x \leq 1, \end{cases}$$

is a maximal dual-feasible function dominating f.

Many standard and simple superadditive functions can be used to generate MDFF using Theorem 2.4. An example is provided and discussed in Sect. 2.4.

2.3.4 Using the Limiting Behaviour of a Function

A maximal dual-feasible function can sometimes be built from a superadditive function $f : [0, 1] \to [0, 1]$ when, at some point x where f is not continuous, the value of $f(x)$ can be increased without modifying the other values. This approach yields an improved function with some singular values x such that

$$\lim_{y \uparrow x} f(y) < f(x) < \lim_{y \downarrow x} f(y).$$

Let $\bar{f}^{x^*}(x) = f(x)$ if $x \neq x^*$, and $\bar{f}^{x^*}(x^*) = f(x^*) + \varepsilon$, with ε being a sufficiently small positive real value. Note that if f is a dual-feasible function, $\bar{f}^{x^*}$ may or may not be a dual-feasible function. Furthermore, for a given dual-feasible function f, let I_1 be the set of values x^* whose images can be increased such that $\bar{f}^{x^*}$ is also a dual-feasible function. Below, we assume that I_1 is a discrete set.

Theorem 2.5 *Let* $f : [0, 1] \to [0, 1]$ *be a superadditive function. Let* I_1 *be the set of values* $x \in [0, 1]$, *for which positive* ε *exist such that* $\bar{f}^x$ *is a dual-feasible function. Assume that* f *is continuous from the right in the entire set* $I_2 := [0, 1] \setminus I_1$ *of the remaining values and that the function* $g : [0, 1] \to [0, 1]$ *is such that the following holds:*

1. $f(x) \leq g(x) \leq \lim_{y \downarrow x} f(y)$, *for any x in I_1;*
2. $g(x) + g(y) \leq g(x + y)$, *if* $x, y, x + y \in I_1$;
3. $g(x) + f(y) \leq g(x + y)$, *if* $x, x + y \in I_1$ *and* $y \in I_2$.

Then the function $h : [0, 1] \to [0, 1]$ *with*

$$h(x) := \begin{cases} g(x), & \text{if } x \in I_1, \\ f(x), & \text{if } x \in I_2, \end{cases}$$

is superadditive.

2.3 Generating One-Dimensional Dual-Feasible Functions

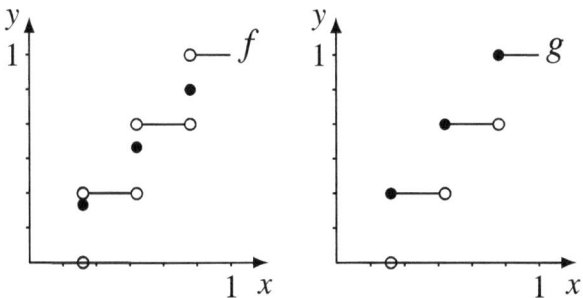

Fig. 2.7 Increasing the values of a function at its discontinuities

Example 2.8 The function $f_{FS,1}(\cdot; k) : [0, 1] \to [0, 1]$, with $k \in \mathbb{N} \setminus \{0\}$, discussed in Sect. 2.3.2 is a MDFF. Hence, the restriction of $f_{FS,1}$ to the domain $[0, C]$ with $\frac{1}{k+1} \leq C < 1$ is superadditive and non-decreasing. Normalizing this function to domain and range $[0, 1]$ yields the following function f, which depends on k and C and which is not necessarily symmetric:

$$f(x) := \begin{cases} Cx/f_{FS,1}(C; k), & \text{if } (k+1) \times Cx \in \mathbb{N}, \\ \lfloor (k+1) \times Cx \rfloor / (k \times f_{FS,1}(C; k)), & \text{otherwise.} \end{cases}$$

This function is continuous on $[0, 1]$ except at the points $\frac{1}{(k+1) \times C}, \frac{2}{(k+1) \times C}, \ldots$ Let for example $k := 5$ and $C := 5/8$, for which $f_{FS,1}(C; k) = 3/5$. Then,

$$f(x) = \begin{cases} 25x/24, & \text{if } 15x/4 \in \mathbb{N}, \\ \lfloor 15x/4 \rfloor / 3, & \text{otherwise,} \end{cases}$$

and $f\left(\frac{4}{15}\right) = \frac{5}{18}, f\left(\frac{8}{15}\right) = \frac{5}{9}$ and $f\left(\frac{4}{5}\right) = \frac{5}{6}$, while f is equal to 0, $\frac{1}{3}$, $\frac{2}{3}$ or 1 in the other points of the interval $[0, 1]$. This allows to set $g(x) := \lfloor 15x/4 \rfloor / 3$ for all $x \in [0, 1]$, and it yields $h \equiv g$. However, this function is not symmetric, as for example $g(1/2) = 1/3 < 1/2$ shows (Fig. 2.7).

Note that to further obtain a maximal dual-feasible function, we may apply Theorem 2.4. □

2.3.5 Rounding Functions

Another way of generating a dual-feasible function consists in applying two superadditive functions separately, one to the integer part of a given value, and the other to its remainder. This approach is valid provided that the conditions of Lemma 2.5 are satisfied.

Lemma 2.5 *Let* $f : [0, C] \to \mathbb{R}_+$ *and* $g : [0, 1] \to [0, 1]$ *be two superadditive functions,* $C \geq 1$ *and*

$$v^* := \sup\{g(y) + g(z) - g(y + z - 1) \mid y, z \in (0, 1] \wedge y + z > 1\}.$$

If

$$f(x + 1) - f(x) \geq v^*$$

for all $x \in [0, C - 1]$, *then the function* $h : [0, C] \to [0, f(\lfloor C \rfloor) + 1]$ *defined by*

$$h(x) := f(\lfloor x \rfloor) + g(\mathrm{frac}(x))$$

is superadditive on $[0, C]$.

Proof Choose any $x, y \in [0, C]$ with $x + y \leq C$. If $\mathrm{frac}(x) + \mathrm{frac}(y) < 1$, then $\mathrm{frac}(x + y) = \mathrm{frac}(x) + \mathrm{frac}(y)$ and $\lfloor x + y \rfloor = \lfloor x \rfloor + \lfloor y \rfloor$, such that $h(x + y) \geq h(x) + h(y)$ follows immediately from the superadditivity of the functions f and g. If $\mathrm{frac}(x) + \mathrm{frac}(y) \geq 1$, then $\mathrm{frac}(x + y) = \mathrm{frac}(x) + \mathrm{frac}(y) - 1$ and $\lfloor x + y \rfloor = \lfloor x \rfloor + \lfloor y \rfloor + 1$. Hence, one gets

$$h(x + y) - h(x) - h(y) = f(\lfloor x + y \rfloor) - f(\lfloor x \rfloor) - f(\lfloor y \rfloor) + g(\mathrm{frac}(x + y))$$
$$- g(\mathrm{frac}(x)) - g(\mathrm{frac}(y))$$
$$\geq f(\lfloor x + y \rfloor) - f(\lfloor x + y \rfloor - 1) - v^*$$
$$\geq 0.$$

□

On the other hand, although the ceiling function is not superadditive, it can lead to superadditive functions if it is decreased by a suitable value. We now generalize several results that use this kind of method.

Lemma 2.6 *Let* $f : \mathbb{R}_+ \to \mathbb{R}_+$ *be a superadditive function. If* $\beta \geq 1$, *then* $g(x) := \max\{0, \lceil f(x) \rceil - \beta\}$ *is superadditive.*

Proof Since f is a superadditive function with domain and range $\mathbb{R}_+$, it is nondecreasing, and hence g is also nondecreasing. Choose any $x, y \geq 0$ with $x \leq y$. If $\lceil f(x) \rceil \leq \beta$, then $g(x) = 0$, such that the superadditivity of g follows from its monotonicity. If $\lceil f(x) \rceil > \beta$, then

$$g(x + y) - g(x) - g(y) = \lceil f(x + y) \rceil - \beta - \lceil f(x) \rceil - \lceil f(y) \rceil + 2\beta$$
$$\geq \beta - 1,$$

since f is superadditive. □

2.4 Examples

In this section, we illustrate the methods discussed above to generate 1-dimensional dual-feasible functions through different examples of functions that have been described in the literature. A particular staircase function is also presented to show the variety of dual-feasible functions that may be further derived.

2.4.1 Applying Symmetry

The function $f_{CCM,1}$ discussed in Example 2.5 (p. 28) can also be obtained by applying Theorem 2.4 to the function $x \mapsto \lfloor Cx \rfloor$, $C \geq 1$:

$$f_{CCM,1}(x; C) := \begin{cases} \lfloor Cx \rfloor / \lfloor C \rfloor, & \text{if } 0 \leq x < 1/2, \\ 1/2, & \text{if } x = 1/2, \\ 1 - f_{CCM,1}(1-x; C), & \text{if } 1/2 < x \leq 1. \end{cases}$$

Rounding-down violates the symmetry, such that the obtained function $x \mapsto \lfloor Cx \rfloor$ is not maximal.

Another example on how applying symmetry may yield a maximal dual-feasible function is described next. The function $g(x)$ described in Sect. 2.1 is a monotone and superadditive dual-feasible function for every $k \in \mathbb{N} \setminus \{0, 1\}$:

$$g(x) := \lceil \max\{0, kx - 1\} \rceil / (k - 1).$$

However, it is not maximal (see for example Fig. 2.2). By forcing symmetry as described in Theorem 2.4, we get the maximal dual-feasible function $f_{VB,2} : [0, 1] \to [0, 1]$ with

$$f_{VB,2}(x; k) := \begin{cases} \lceil \max\{0, kx-1\} \rceil / (k-1), & \text{if } 0 \leq x < 1/2, \\ 1/2, & \text{if } x = 1/2, \\ 1 - f_{VB,2}(1-x; k), & \text{if } 1/2 < x \leq 1. \end{cases} \quad (2.11)$$

Figure 2.8 illustrates this function for $k = 6$.

2.4.2 Using Rounding Functions and Applying Symmetry

In this subsection, we assume $C > 1$ and that C is not integer. Let k be an integer constant with $k \geq \left\lceil \frac{1}{\text{frac}(C)} \right\rceil$. The following function $f_{LL,1} : [0, 1] \to [0, 1]$ is based

Fig. 2.8 Function $f_{VB,2}$ for $k = 6$

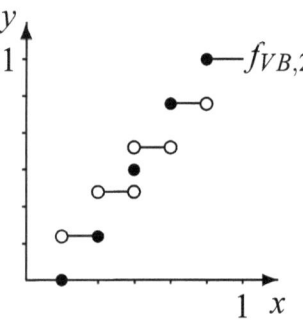

on Lemma 2.5:

$$f_{LL,1}(x; C, k) := \frac{\lfloor Cx \rfloor + \max\left\{0, \left\lceil \frac{\text{frac}(Cx) - \text{frac}(C)}{1 - \text{frac}(C)} \times (k-1) \right\rceil / k\right\}}{\lfloor C \rfloor}. \quad (2.12)$$

The superadditivity of this function is due to Lemma 2.6. In the following proof, we use the fact that $\lceil x+y \rceil = x + \lceil y \rceil$ if x is integer, and that $\lceil x \rceil + \lceil y \rceil \leq \lceil x+y \rceil + 1$ for any x and y. Without the maximum expression, the function $f_{LL,1}$ would be the function (2.1). The function $f_{LL,1}$ is derived from another superadditive function h by a linear transformation, namely $f_{LL,1}(x) = h(Cx)/\lfloor C \rfloor$, where the structure of h is of the kind $f(\lfloor x \rfloor) + g(\text{frac}(x))$ as in Lemma 2.5.

Proposition 2.5 *Function $f_{LL,1}$ is superadditive.*

Proof To show the superadditivity of $f_{LL,1}$, Lemmas 2.5 and 2.6 are used. Define the function $g : [0, 1] \to [0, 1]$ as

$$g(x) := \max\left\{0, \left\lceil \frac{x - \text{frac}(C)}{1 - \text{frac}(C)} \times (k-1) \right\rceil\right\} / k.$$

The range of g is indeed part of $[0, 1]$, because $x \in [0, 1]$ implies $\frac{x - \text{frac}(C)}{1 - \text{frac}(C)} \leq 1$. In the following, the superadditivity of g is proved. The function $x \mapsto \frac{x}{1 - \text{frac}(C)} \times (k-1)$ is linear, and hence it is obviously superadditive. The constant $\frac{\text{frac}(C)}{1 - \text{frac}(C)} \times (k-1)$ is at least one, because

$$k - 1 \geq \left\lceil \frac{1}{\text{frac}(C)} - 1 \right\rceil \geq \frac{1 - \text{frac}(C)}{\text{frac}(C)}.$$

If f is any superadditive function and $\alpha \geq 0$, then $x \mapsto f(x) - \alpha$ is also superadditive. Hence, Lemma 2.6 can be applied, even if the expression $\frac{-\text{frac}(C)}{1 - \text{frac}(C)} \times (k-1)$ stands inside the rounding brackets. The function g is therefore superadditive.

2.4 Examples

The identity function is superadditive. To apply Lemma 2.5 with the above defined function g, we must show that $v^* \leq 1$. Choose any $y, z \in (0, 1]$ with $y + z > 1$. Since $g(y), g(z)$ and $g(y + z - 1) \in [0, 1]$, the inequality

$$g(y) + g(z) - g(y + z - 1) \leq 1$$

is obviously fulfilled for $g(y) \times g(z) = 0$. Assume $g(y)$ and $g(z) > 0$. Hence, we have y and $z > \text{frac}(C)$. One gets

$$\begin{aligned}
k \times g(y + z - 1) &\geq \left\lceil \frac{y + z - 1 - \text{frac}(C)}{1 - \text{frac}(C)} \times (k - 1) \right\rceil \\
&= \left\lceil \left(\frac{y - \text{frac}(C) + z - \text{frac}(C)}{1 - \text{frac}(C)} - 1 \right) \times (k - 1) \right\rceil \\
&\geq \left\lceil \frac{y - \text{frac}(C)}{1 - \text{frac}(C)} \times (k - 1) \right\rceil + \left\lceil \frac{z - \text{frac}(C)}{1 - \text{frac}(C)} \times (k - 1) \right\rceil - k \\
&= k \times (g(y) + g(z) - 1),
\end{aligned}$$

as needed, because $k \in \mathbb{N}$.

Finally, $f_{LL,1}(x) = h(Cx)/\lfloor C \rfloor$ with the function h according to Lemma 2.5. The linear transformations do not affect the superadditivity. Since h is superadditive, $f_{LL,1}$ is that too. $\square$

The function $f_{LL,1}$ is not maximal, since there are cases where it is not symmetric. An improved version of this function can be obtained by applying Theorem 2.4.

Proposition 2.6 *The following function $f_{LL,2}(\cdot; C, k) : [0, 1] \to [0, 1]$ with $C > 1$, $C \notin \mathbb{N}$ and $k \in \mathbb{N}$, $k \geq \lceil 1/\text{frac}(C) \rceil$ is a maximal dual-feasible function, and it dominates $f_{LL,1}$.*

$$f_{LL,2}(x; C, k) := \begin{cases} \frac{\lfloor Cx \rfloor + \max\left\{0, \left\lceil \frac{\text{frac}(Cx) - \text{frac}(C)}{1 - \text{frac}(C)} \times (k-1) \right\rceil / k \right\}}{\lfloor C \rfloor}, & \text{if } 0 \leq x < 1/2, \\ 1/2, & \text{if } x = 1/2, \\ 1 - f_{LL,2}(1 - x; C, k), & \text{if } 1/2 < x \leq 1. \end{cases} \quad (2.13)$$

The graphs of the functions (2.12) and (2.13) are drawn in Fig. 2.9 for $C \in \{3.3, 3.4\}$ and $k = \lceil 1/\text{frac}(C) \rceil$.

2.4.3 Improving a Function by Using Its Limiting Behaviour

The following four parameter dependent functions $u_A, u_B, u_C, u_D : [0, 1] \to [0, 1]$ are among the first functions described explicitly in the literature as being

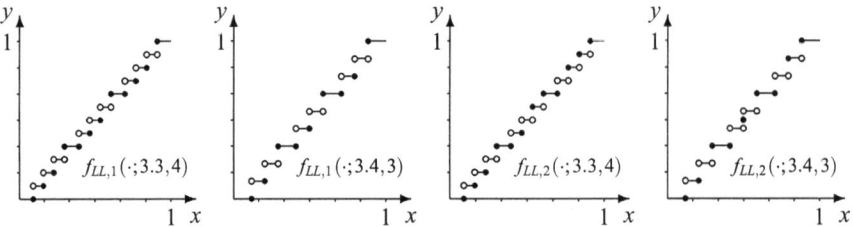

Fig. 2.9 Original $f_{LL,1}(\cdot;C,k)$ and improved $f_{LL,2}(\cdot;C,k)$

dual-feasible functions:

$$u_A(x;p) := \begin{cases} 0, & \text{if } 0 \le x \le \frac{1}{p+1}, \\ \frac{1}{p}, & \text{if } \frac{1}{p+1} < x \le 1, \end{cases} \quad p \in \mathbb{N} \setminus \{0\},$$

$$u_B(x;p,a) := \begin{cases} 0, & \text{if } 0 \le x \le ap - \frac{p-1}{p+1}, \\ \frac{1}{p+1} + (x - \frac{1}{p+1})/(p - ap - ap^2), & \text{if } ap - \frac{p-1}{p+1} < x < \beta, \\ \frac{1}{p}, & \text{if } \beta \le x \le 1, \end{cases}$$

$$p \in \mathbb{N} \setminus \{0,1\}, \ a \in (\frac{p-1}{p^2+p}, \frac{1}{p+1}),$$

$$u_C(x;p) := \max\{0, \lceil (p+1)x - 1 \rceil / p\}, \quad p \in \mathbb{N} \setminus \{0\},$$

$$u_D(x;p,a) := \left\lfloor \frac{x}{\beta} \right\rfloor / p + u_B(x - \beta \times \left\lfloor \frac{x}{\beta} \right\rfloor; p, a),$$

$$p \in \mathbb{N} \setminus \{0,1,2\}, \ a \in (\frac{p-1}{p^2+p}, \frac{1}{p+1}),$$

where $\beta := \frac{2}{p+1} - a$ for the functions u_B and u_D. The function u_C is equivalent to the function (2.2) with $k = p+1$. Furthermore, we have for all $x \in [0, \frac{2}{p+1}]$ that $u_A(x) = u_C(x)$. Similarly, $u_B(x) = u_D(x)$ holds for all $x \in [0, \beta]$. For larger x inside the interval $[0,1]$ one gets $u_C(x) > u_A(x)$ or $u_D(x) > u_B(x)$, respectively. The function u_A is a simple staircase function. The dependence of u_B and u_D on the parameter a is shown in Fig. 2.10 for $p = 3$. Since u_D is continuous and piecewise linear, it looks like the function (2.4) (p. 25) for $C := 1/\beta$. Checking this observation is left as an exercise.

The function $f_{FS,1} : [0,1] \to [0,1]$ depending on the parameter $k \in \mathbb{N} \setminus \{0\}$ discussed in Sect. 2.3.2:

$$f_{FS,1}(x;k) := \begin{cases} x, & \text{if } (k+1) \times x \in \mathbb{N}, \\ \lfloor (k+1) \times x \rfloor / k, & \text{otherwise}, \end{cases}$$

is a maximal dual-feasible function. This function was obtained from u_C by applying Theorem 2.5. The values that are equal to $1/(k+1), 2/(k+1), \ldots, k/(k+1)$ remain unchanged, while the other ones are submitted to a rounding function.

2.4 Examples 43

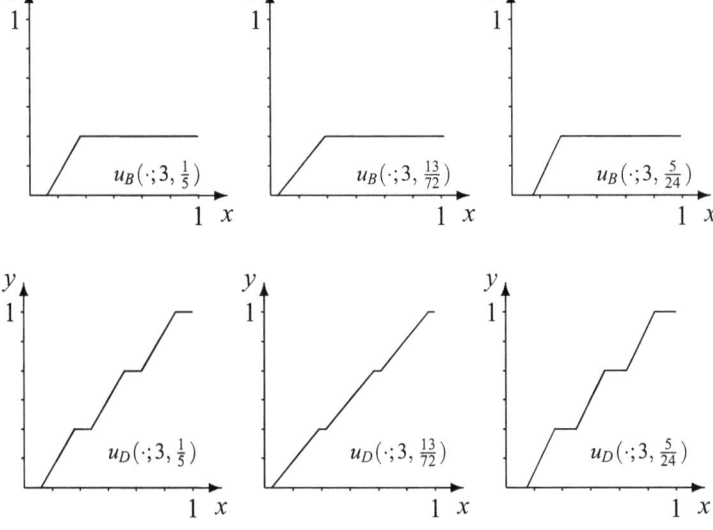

Fig. 2.10 Dual-feasible functions u_B and u_D for $p = 3$ and $a \in \{\frac{1}{5}, \frac{13}{72}, \frac{5}{24}\}$

The following function $f_{DG,1}$ is also built on Theorem 2.5. Since the proof of superadditivity is complex and long, it is omitted.

Proposition 2.7 *For every $C \in \mathbb{R} \setminus \mathbb{N}$ with $C > 1$, and any $k \in \mathbb{N}$ with $k \geq \lceil \frac{1}{\text{frac}(C)} \rceil$, the following function $f_{DG,1}(\cdot; C, k) : [0, 1] \to [0, 1]$ is a MDFF:*

$$f_{DG,1}(x) = \frac{\lfloor Cx \rfloor}{\lfloor C \rfloor} + \frac{1}{\lfloor C \rfloor} \times \begin{cases} \frac{\text{frac}(Cx)-\text{frac}(C)}{1-\text{frac}(C)}, & \text{if } (k-1) \times \frac{\text{frac}(Cx)-\text{frac}(C)}{1-\text{frac}(C)} \in \mathbb{N}, \\ \max\left\{0, \left\lceil (k-1) \times \frac{\text{frac}(Cx)-\text{frac}(C)}{1-\text{frac}(C)} \right\rceil / k \right\}, & \text{otherwise.} \end{cases}$$
(2.14)

This function also dominates $f_{LL,1}$, if $f_{LL,1}$ is not symmetric. In this case, the functions (2.12), (2.13) and (2.14) differ at some isolated points. The graph of the latter function is presented for $C \in \{3.3, 3.4\}$ and $k = \lceil 1/\text{frac}(C) \rceil$ in Fig. 2.11.

2.4.4 A Special Case: A Staircase Function with Infinitely Many Stairs

All maximal dual-feasible functions that were presented until now have a relatively simple structure. For instance, the function $f_{BJ,1}$ is Lipschitz-continuous, while all the other discussed dual-feasible functions have a finite number of discontinuities.

Fig. 2.11 Function $f_{DG,1}$

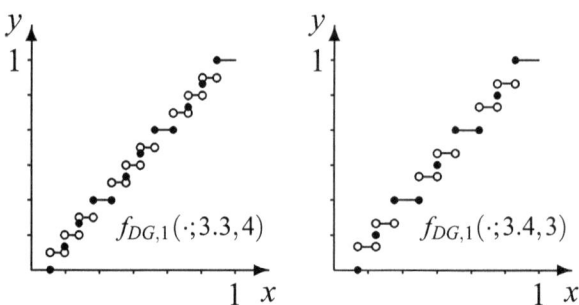

$f_{DG,1}(\cdot;3.3,4)$ $f_{DG,1}(\cdot;3.4,3)$

The following example illustrates that maximal dual-feasible functions may be much more complicated.

Although it has an infinite number of stairs, the function $f : [0,1] \to [0,1]$ defined as follows

$$f(x) := \begin{cases} \frac{1}{2} + 1/\lfloor \frac{2}{2x-1} \rfloor, & \text{for } \frac{1}{2} < x \leq 1, \\ 1/2, & \text{for } x = 1/2, \\ \frac{1}{2} - 1/\lfloor \frac{2}{1-2x} \rfloor, & \text{for } 0 \leq x < \frac{1}{2}, \end{cases} \quad (2.15)$$

is a maximal dual-feasible function, which is differentiable at $x_0 = 1/2$ with the derivative $f'(1/2) = 1$.

Proof Clearly, f is monotonically increasing and symmetric in $[0, 1]$, and it has for all $n \in \mathbb{N}$ with $n > 2$ discontinuities at $\frac{1}{2} \pm \frac{1}{n}$. The stairs have the levels 0, 1 and $\frac{1}{2} \pm \frac{1}{n}$. Furthermore, $f(x) = 0$ for $0 \leq x < 1/6$. To show the superadditivity, a small case distinction is necessary. According to Theorem 2.2, assume $0 < x_1 \leq x_2 < 1/2$ and $x_1 + x_2 \leq 2/3$.

- If $x_1 + x_2 \geq 1/2$, then $f(x_1 + x_2) \geq x_1 + x_2 \geq f(x_1) + f(x_2)$, because $f(x) \leq x$ for $0 \leq x \leq 1/2$.
- If $0 \leq x_1 < 1/6$, then $f(x_1) = 0$, and therefore, $f(x_1 + x_2) \geq f(x_1) + f(x_2)$.
- Let $1/6 \leq x_1 \leq x_2$ and $x_1 + x_2 < 1/2$. Then, $x_2 < 1/3$ and $x_1 < 1/4$, i.e., $f(x_1) = 1/6$.

  - If $x_2 < 1/4$, then $f(x_2) = 1/6$ and $f(x_1 + x_2) \geq f(2 \times 1/6) = 1/3 = f(x_1) + f(x_2)$.
  - If $1/4 \leq x_2 < 3/10$, then $f(x_2) = 1/4$ and $f(x_1 + x_2) \geq f(5/12) = 5/12 = f(x_1) + f(x_2)$.
  - If $3/10 \leq x_2 < 1/3$, then $f(x_2) = 3/10$ and $f(x_1 + x_2) \geq f(7/15) = 7/15 = f(x_1) + f(x_2)$.

Fig. 2.12 Maximal dual-feasible staircase function (2.15) with infinitely many stairs

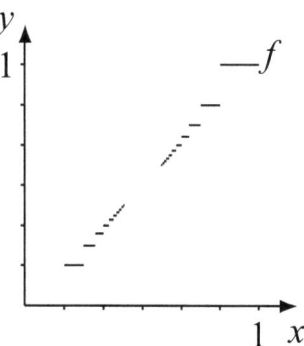

To show that f is differentiable with derivative 1 at $x_0 = 1/2$, let $x = 1/2 + 1/(a+b)$ with $a \in \mathbb{N}$, $a \geq 2$ and $0 \leq b < 1$. (The case $x < 1/2$ can be handled analogously or by use of the symmetry.) Then,

$$f(x) - f(x_0) - 1 \times (x - x_0) = \frac{1}{\lfloor \frac{2}{1 + \frac{2}{a+b} - 1} \rfloor} - \frac{1}{a+b}$$

$$= \frac{1}{\lfloor a+b \rfloor} - \frac{1}{a+b}$$

$$= \frac{1}{a} - \frac{1}{a+b}$$

$$= \frac{b}{a^2 + ab}.$$

Hence, the fraction is nonnegative. When x tends to $1/2$, then $a \to \infty$, and therefore $f(x) - f(x_0) - 1 \times (x - x_0) = \mathcal{O}(x - x_0)^2$. This means that the derivative at $x_0 = 1/2$ exists and has the value 1. □

Figure 2.12 shows a simplified version of this staircase function. Around the point $(1/2; 1/2)$, there are infinitely many stairs, which cannot be drawn exactly. Moreover, the usual black and white marks at discontinuities become meaningless, and therefore, they are omitted.

2.5 Related Literature

The concept of 1-dimensional dual-feasible function was described first by Johnson (1973). In the same year, a subclass of dual-feasible functions restricted to super-additive and nondecreasing functions was described to strengthen cuts for integer programs by Chvátal (1973) [see also Aardal and Weismantel (1997), Nemhauser and Wolsey (1998)]. Lueker (1983) introduced the designation. He used the dual-feasible functions to derive lower bounds for bin-packing problems. The functions $u_A(\cdot;p)$, $u_B(\cdot;p,a)$, $u_C(\cdot;p)$ and $u_D(\cdot;p,a)$ described in Sect. 2.4.3 are due to this author. The notion of maximality was proposed and described initially in Carlier and Néron (2007a), while the conditions for maximality were presented by these authors in Carlier and Néron (2007b). In the latter, the authors defined also the discrete version of a dual-feasible function that they call *redundant functions*. They used these functions to solve scheduling problems. The conditions for maximality were later reviewed by Rietz et al. (2010) yielding the sufficient conditions described in Theorem 2.2. The proof of Theorem 2.4 can be found in Clautiaux et al. (2010) for the case where the function is defined as a discrete dual-feasible function. The extremality of dual-feasible functions was analyzed first by Rietz et al. (2012a). The proofs of many of the results concerning extremality can be found in this reference.

Within this chapter, different dual-feasible functions were used as examples. Most of them were taken from the literature where they were stated either explicitly or implicitly. The letters used in the index of these functions identify the authors that proposed them originally. We provide the original source next. The function $f_{BJ,1}(\cdot;C)$ is based on the work of Burdett and Johnson (1977). The function $f_{CCM,1}(\cdot;C)$ was proposed by Carlier et al. (2007), improving slightly a function proposed before by Boschetti and Mingozzi (2003). The function $f_{FS,1}(\cdot;k)$ is due to Fekete and Schepers (2001). The function $f_{VB,2}(\cdot;k)$ is based on a non-maximal dual-feasible function by Vanderbeck (2000). The function $f_{VB,2}(\cdot;k)$ is maximal, and it was described first by Clautiaux et al. (2010). The function $f_{DG,1}(\cdot;k)$ was defined by Dash and Günlük (2006). The function $f_{LL,1}(\cdot;C,k)$ was used implicitly by Letchford and Lodi (2002) to strengthen Chvátal-Gomory cuts (Chvátal 1973) and Gomory fractional cuts (Gomory 1958) in linear programs. As shown in Sect. 2.4.2, this function is superadditive but not maximal. Again, the corresponding maximal dual-feasible function was defined by Clautiaux et al. (2010).

2.6 Exercises

1. Which of the following functions $f_1, f_2 : [0, 1] \to [0, 1]$ are dual-feasible?

$$f_1(x) := \begin{cases} x, & \text{if } x \in \mathbb{Q}, \\ 1/\lceil 1/x \rceil, & \text{otherwise,} \end{cases}$$

$$f_2(x) := \begin{cases} 0, & \text{if } x = 0, \\ 1/\lfloor 1/x \rfloor, & \text{otherwise.} \end{cases}$$

2. Show that if $f : \mathbb{R} \to \mathbb{R}$ is superadditive, then $\lfloor f \rfloor$ remains superadditive.

2.6 Exercises

3. Which of the following functions $f_3, f_4 : \mathbb{R}_+ \to \mathbb{R}_+$ are superadditive?

$$f_3(x) := \lfloor x^2 \rfloor,$$

$$f_4(x) := \lfloor e^x \rfloor.$$

4. Let $g_0 : \{0, 1, \ldots, 19\} \to \{0, 1, \ldots, 10\}$ be the following discrete function:

x	0	1	2	3	4	5	6	7	8	9	10	11	12	13	14	15	16	17	18	19
$g_0(x)$	0	0	0	0	0	2	3	4	5	0	0	0	0	0	0	0	0	0	0	10

   (a) Provide a discrete dual-feasible function $g_1 : \{0, 1, \ldots, 19\} \to \{0, 1, \ldots, 10\}$ with $g_1(15) = 9$, which dominates g_0 and is maximal.
   (b) What is the equivalent dual-feasible function $f_5 : [0, 1] \to [0, 1]$ to g_1 according to (2.3)?
   (c) Provide a continuous, piecewise linear maximal dual-feasible function $f_6 : [0, 1] \to [0, 1]$, which dominates f_5.
   (d) Define a staircase maximal dual-feasible function $f_7 : [0, 1] \to [0, 1]$, which dominates f_5.

5. Consider an instance of the 1D-CSP with given nonnegative integer data $m, L, \mathbf{l}, \mathbf{b}$, where $0 < \ell_i \leq L$ for $i = 1, \ldots, m$. Which of the following statements are true, and which are not?
   Hint: The functions $f_{FS,1}, f_{VB,2}, f_{BJ,1}$ *and* $f_{CCM,1}$ *were defined in (2.4), (2.9), (2.10), (2.11) (pp. 25, 28, 35, 39, respectively).*

   (a) If $f : [0, 1] \to [0, 1]$ is a maximal dual-feasible function, then there is no other dual-feasible function $g : [0, 1] \to [0, 1]$ with $\sum_{i=1}^{m} b_i \times f(\ell_i/L) < \sum_{i=1}^{m} b_i \times g(\ell_i/L)$.
   (b) For all $k \in \mathbb{N} \setminus \{0\}$, there is always a $C \geq 1$, $C \in \mathbb{R}$ with $f_{FS,1}(\ell_i/L; k) = f_{BJ,1}(\ell_i/L; C)$ for $i = 1, \ldots, m$.
   (c) For all $k \in \mathbb{N} \setminus \{0, 1\}$, there is always a $C \geq 1$, $C \in \mathbb{R}$ with $f_{VB,2}(\ell_i/L; k) = f_{CCM,1}(\ell_i/L; C)$ for $i = 1, \ldots, m$.
   (d) For all $k \in \mathbb{N} \setminus \{0, 1\}$, there is always a $C \geq 1$, $C \in \mathbb{R}$ with $f_{VB,2}(\ell_i/L; k) = f_{BJ,1}(\ell_i/L; C)$ for $i = 1, \ldots, m$.

6. Show, using Theorem 2.2, that $f_{BJ,1}(x; C)$ defined in (2.4) is maximal for all $C \geq 1$.

7. Let $p \in \mathbb{R}$ be a constant with $p > 1$. Is the function $f : [0, 1] \to [0, 1]$ with

$$f(x) := \begin{cases} (2x)^p/2, & \text{if } x \leq 1/2, \\ 1 - (2 - 2x)^p/2, & \text{otherwise,} \end{cases}$$

a maximal dual-feasible function? Is it an extreme maximal dual-feasible function?

Hint: Refer to Lemma 2.3 and Theorem 2.3.

8. Prove that $f_{FS,1}(x;k)$ is extreme.

9. Which of the following statements are true and which are not?

    (a) The composition $f(g(\cdot))$ of two maximal dual-feasible functions $f, g : [0, 1] \to [0, 1]$ is always a non-extreme maximal dual-feasible function if f is non-extreme and g is surjective, i.e., a mapping onto and not only into $[0, 1]$.

    (b) A maximal dual-feasible function $f : [0, 1] \to [0, 1]$ is surjective if and only if it is continuous.

    (c) If an extreme maximal dual-feasible function $f : [0, 1] \to [0, 1]$ is convex on $[0, 1/2]$, then f is continuous on $[0, 1]$.

10. Let $\lambda \in (0, 1/2]$ be a fixed parameter for the function $f_{MT,0} : [0, 1] \to [0, 1]$, defined by

$$f_{MT,0}(x; \lambda) := \begin{cases} 0, & \text{if } x < \lambda, \\ 1, & \text{if } x > 1 - \lambda, \\ x, & \text{otherwise.} \end{cases} \quad (2.16)$$

    (a) Show that this function is a maximal dual-feasible function.
    (b) Show that this function is extreme for $\lambda \leq 1/4$.
    Hint: Refer to Lemma 2.4.

11. Show according to Definition 2.6 (p. 28) that if one has the constants $\alpha, \beta \in \mathbb{R}_+$ with $\alpha\beta > 0$ and two different maximal dual-feasible functions $f, g : [0, 1] \to [0, 1]$, then $h := \frac{\alpha}{\alpha+\beta} f + \frac{\beta}{\alpha+\beta} g$ is a non-extreme maximal dual-feasible function.

12. Give an example that the composition of two non-symmetric, non-superadditive dual-feasible functions $f, g : [0, 1] \to [0, 1]$ may yield a maximal dual-feasible function.

13. The function $f_{LL1} : [0, 1] \to [0, 1]$ was defined in (2.12) as

$$f_{LL,1}(x; C, k) := \left(\lfloor Cx \rfloor + \max\left\{ 0, \left\lceil \frac{\text{frac}(Cx) - \text{frac}(C)}{1 - \text{frac}(C)} \times (k-1) \right\rceil / k \right\} \right) / \lfloor C \rfloor.$$

For which parameter values $C > 1$, $C \in \mathbb{R} \setminus \mathbb{N}$ and $k \in \mathbb{N}$, $k \geq \lceil 1/\text{frac}(C) \rceil$ is this function symmetric?

14. Show that the function $f_{VB,1} : [0, 1] \to [0, 1]$, defined in (2.2) as

$$f_{VB,1}(x; k) := \lceil \max\{0, kx - 1\} \rceil / (k - 1),$$

with $k \in \mathbb{N} \setminus \{0, 1\}$, is not symmetric for any parameter value k.

2.6 Exercises

15. Let $C \in \mathbb{R}$ with $C \geq 1$ be a constant. Which of the following statements are true and which are not?

(a) For every maximal dual-feasible function $g : [0, 1] \to [0, 1]$, the function $f : [0, 1] \to \mathbb{R}_+$, defined as $f(x) := \lfloor Cx \rfloor + g(\text{frac}(Cx))$, is superadditive.

(b) For every superadditive function $g : \mathbb{R} \to \mathbb{R}$, the function $f : \mathbb{R} \to \mathbb{R}_+$, defined as $f(x) := \max\{0, \lceil g(x) \rceil - C\}$, is superadditive.

(c) If $C \in \mathbb{N}$, then for every dual-feasible function $f : [0, 1] \to [0, 1]$, and any finite index set I of nonnegative real numbers x_i with $i \in I$, the following implication holds:

$$\sum_{i \in I} x_i \leq \frac{1}{C} \implies \sum_{i \in I} f(x_i) \leq \frac{1}{C}.$$

Chapter 3
General Dual-Feasible Functions

3.1 Introduction

Classical dual-feasible functions are defined only for nonnegative arguments thus limiting their applicability. In this chapter, we explore the extension of dual-feasible functions to more general domains with a focus on real numbers. Other attempts of generalizing the concept of dual-feasible function will be done later in the book. In Chap. 4, we will discuss for instance an extension to multidimensional domains yielding the so-called *vector packing dual-feasible functions*, which may be used to compute bounds for vector packing problems.

Extending the principles of dual-feasible functions to the domain of real numbers is not trivial. The properties that apply to dual-feasible functions, and which have been reviewed in the previous chapter, are affected in this exercise, and some of them are even lost. This makes the task of deriving good non-dominated functions much more difficult. In the sequel, we will explore in depth the new properties of general dual-feasible functions. Different examples will be brought to discussion to illustrate the main and new ideas behind these functions.

Given the hardness in deriving dual-feasible functions that apply to the domain of real numbers, we will devote the second part of the chapter to the presentation of general construction principles that lead to specific instances of general dual-feasible functions. The defining characteristics of these principles will be described first, and followed by the analysis of specific examples for each case.

3.2 Extension of Dual-Feasible Functions to General Domains

3.2.1 Definition

A general dual-feasible function is a generalization of the classical dual-feasible functions to any real arguments, i.e. arguments that are not restricted to nonnegative values. The definition of a general dual-feasible function states formally as follows.

Definition 3.1 A function $f : \mathbb{R} \to \mathbb{R}$ is a *general dual-feasible function*, if for any finite index set I of real numbers $x_i \in \mathbb{R}$, $i \in I$, it holds that

$$\sum_{i \in I} x_i \leq 1 \implies \sum_{i \in I} f(x_i) \leq 1. \tag{3.1}$$

Extending the domain and range of dual-feasible functions to $\mathbb{R}$ instead of $[0, 1]$ has several facets. As we will see, this extension is far from trivial, because important known properties of the bounded classical case are lost. For instance, any maximal dual-feasible function with domain and range $[0, 1]$ obeys a certain symmetry rule. It was possible to characterize these functions by an equivalence, i.e. they were maximal dual-feasible functions if and only if they fulfilled a certain set of rules. However, while we extend the domain and range to the field of real numbers, the mentioned symmetry rule is not longer a necessary property of the maximal dual-feasible functions. Some similar rules as in the classical case remain sufficient, but it is not longer possible to formulate an analogous equivalence in the generalized case.

Example 3.1 The classical dual-feasible function $f_{BJ,1} : [0, 1] \to [0, 1]$ was defined in Chap. 2 for any real parameter $C \geq 1$ as

$$f_{BJ,1}(x; C) := \left(\lfloor Cx \rfloor + \max \left\{ 0, \frac{\mathsf{frac}(Cx) - \mathsf{frac}(C)}{1 - \mathsf{frac}(C)} \right\} \right) / \lfloor C \rfloor,$$

where $\mathsf{frac}(\cdot)$ denotes the non-integer part of an expression, i.e. $\mathsf{frac}(C) \equiv C - \lfloor C \rfloor$. This function can be extended to a general dual-feasible function using the same formula. Note however that this does not work for many other functions. Figure 3.1 illustrates this function $f_{BJ,1}$ for $C \in \{6/5, 8/5\}$.

□

Further conditions for a function $f : \mathbb{R} \to \mathbb{R}$ to be a general dual-feasible function have already been identified. The following proposition describes two of them. The proof of their validity is left as an exercise (see Exercise 1 at the end of the chapter).

Proposition 3.1 *Let $f : \mathbb{R} \to \mathbb{R}$ be any function. If f is a general dual-feasible function, then it has the following two properties.*

3.2 Extension of Dual-Feasible Functions to General Domains

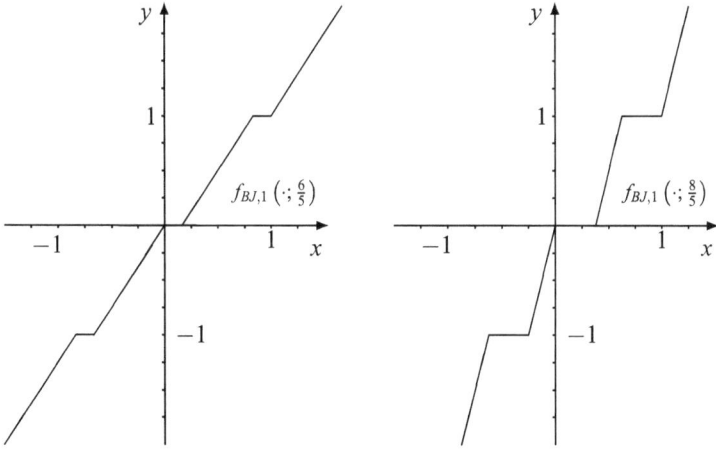

Fig. 3.1 $f_{BJ,1}$ as a maximal general dual-feasible function

1. For any $x \in (0, 1]$, it holds that $f(x) \leq 1/\lfloor 1/x \rfloor$.
2. For any finite set $\{x_i \in \mathbb{R} : i \in I\}$ of real numbers, the following holds:

$$\sum_{i \in I} x_i \leq 0 \implies \sum_{i \in I} f(x_i) \leq 0 \qquad (3.2)$$

The converse is generally false.

These properties are necessary but not sufficient as the following example illustrates.

Example 3.2 Let $c > 1$ be any real constant. The following function f shows that even when the properties (1.) and (2.) in Proposition 3.1 are fulfilled, the function may not be a general dual-feasible function:

$$f(x) := \begin{cases} x, & \text{if } 0 \leq x \leq 1, \\ cx, & \text{otherwise.} \end{cases}$$

The property (1.) is obviously fulfilled. The same happens with (2.) because $f(x) \leq cx$ for all $x \in \mathbb{R}$. However, we have that $f(-1) + f(2) = c > 1$, in contradiction to $-1 + 2 \leq 1$, and thus a violation of the defining condition (3.1). □

On the other hand, the composition of general dual-feasible functions leads to functions that are still general dual-feasible functions.

Lemma 3.1 *The composition of general dual-feasible functions $f, g : \mathbb{R} \to \mathbb{R}$ is a general dual-feasible function.*

Proof Let I be any finite index set of real numbers x_i with $\sum_{i \in I} x_i \leq 1$. One gets $\sum_{i \in I} g(x_i) \leq 1$, because g is a general dual-feasible function. Therefore, Definition 3.1

yields
$$\sum_{i \in I} f(g(x_i)) \leq 1,$$
because f is also a general dual-feasible function. □

3.2.2 Maximality

As happens with classical dual-feasible functions, only non-dominated functions are of interest. In this section, similarly to the bounded case, the notion of *maximal general dual-feasible function* is introduced. We will also provide characterising conditions for these functions and point out the possible loss of symmetry.

Definition 3.2 A general dual-feasible function f is *maximal*, if there is no other general dual-feasible function g with $f(x) \leq g(x)$ for all $x \in \mathbb{R}$.

The simplest maximal general dual-feasible functions are the linear ones according to the following proposition, but even these functions raise some exceptions.

Proposition 3.2 *For every $c \in [0, 1]$, the linear function $f : \mathbb{R} \to \mathbb{R}$ with $f(x) := cx$ is a maximal general dual-feasible function.*

Proof The function f is a general dual-feasible function according to Definition 3.1, because for any $n \in \mathbb{N} \setminus \{0\}$ and numbers $x_1, \ldots, x_n \in \mathbb{R}$ with $\sum_{i=1}^{n} x_i \leq 1$, it holds that
$$\sum_{i=1}^{n} f(x_i) = c \times \sum_{i=1}^{n} x_i \leq c.$$

Suppose there is a general dual-feasible function $g : \mathbb{R} \to \mathbb{R}$ with $g(x) \geq cx$ for all $x \in \mathbb{R}$, and $g(y) > cy$ for a certain $y \in \mathbb{R}$. Definition 3.1 implies $g(y) + g(-y) \leq 0$. Since $g(-y) \geq f(-y)$, the contradiction
$$0 \geq g(y) + g(-y) > cy - cy = 0$$
follows. Since f is not dominated by another general dual-feasible function, it is maximal. □

The previous proposition shows that, in the case of general dual-feasible functions, the symmetry rule
$$f(x) + f(1 - x) = 1, \quad \text{for all} \quad x \leq 1/2, \tag{3.3}$$

3.2 Extension of Dual-Feasible Functions to General Domains

is no more a necessary condition for the function to be maximal. The conditions for maximality are restated as follows.

Theorem 3.1 *Let $f : \mathbb{R} \to \mathbb{R}$ be a given function.*

(a) *If f satisfies the following conditions, then f is a general MDFF:*

1. $f(0) = 0$;
2. f *is superadditive, i.e. for all $x, y \in \mathbb{R}$, it holds that*

$$f(x+y) \geq f(x) + f(y); \tag{3.4}$$

3. *there is an $\varepsilon > 0$, such that $f(x) \geq 0$ for all $x \in (0, \varepsilon)$;*
4. *f obeys the symmetry rule (3.3);*

(b) *If f is a general MDFF, then the above properties (1.)–(3.) hold for f, but not necessarily (4.);*
(c) *If f satisfies the above conditions (1.)–(3.), then f is monotonely increasing;*
(d) *If the symmetry rule (3.3) holds and f obeys the inequality (3.4) for all $x, y \in \mathbb{R}$ with $x \leq y \leq \frac{1-x}{2}$, then f is superadditive.*

Unlike for classical dual-feasible functions, where the range was [0, 1], here the nonnegativity of the function values for nonnegative arguments must explicitly be demanded.

Example 3.3 The following example of a continuous and piecewise linear function $f : \mathbb{R} \to \mathbb{R}$ (Fig. 3.2) shows that Theorem 3.1 would become false, if the prerequisite (3.) in part (a) was not considered.

$$f(x) := \begin{cases} 5x, & \text{if } x \leq 0, \\ -x, & \text{if } 0 \leq x \leq 1/3, \\ 5x - 2, & \text{if } 1/3 \leq x \leq 2/3, \\ 2 - x, & \text{if } 2/3 \leq x \leq 1, \\ 5x - 4, & \text{if } x \geq 1. \end{cases}$$

The conditions (1.) and (4.) in Theorem 3.1 can easily be checked. To verify the superadditivity, choose any $x, y \in \mathbb{R}$ with

$$x \leq y \leq \frac{1-x}{2}$$

according to part (d) of Theorem 3.1. Hence, we have $x \leq 1/3$ and

$$x + y \leq \frac{1+x}{2} \leq \frac{2}{3}.$$

Let $d := f(x+y) - f(x) - f(y)$. A small case distinction is necessary to show $d \geq 0$.

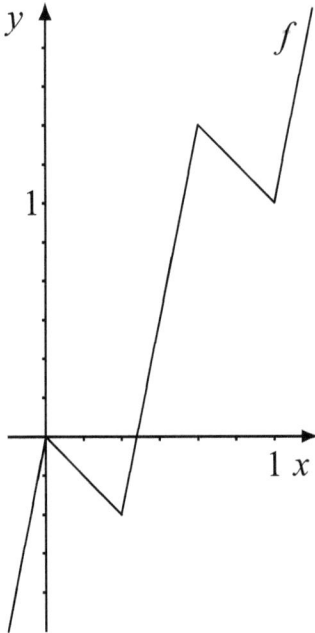

Fig. 3.2 Necessity of the monotonicity

1. $x > 0$ yields $f(x) = -x$ and $0 < y < 1/2$. If $y \leq 1/3$, then $f(y) = -y$ and
$$f(x+y) \geq -x - y,$$
and hence $d \geq 0$. If $y > 1/3$, then $f(y) = 5y - 2$ and
$$f(x+y) = 5x + 5y - 2,$$
and hence $d = 6x > 0$.
2. $x \leq 0$ yields $f(x) = 5x$, and therefore
$$d = f(x+y) - f(y) - 5x \geq 0.$$

The details of the proof in this case are omitted. □

3.2.3 Extremality

The notion of extremality remains essentially the same as for classical dual-feasible functions.

3.2 Extension of Dual-Feasible Functions to General Domains

Definition 3.3 A maximal general dual-feasible function $f : \mathbb{R} \to \mathbb{R}$ is *extreme*, if any maximal general dual-feasible functions $g, h : \mathbb{R} \to \mathbb{R}$ with $2f(x) = g(x) + h(x)$ for all $x \in \mathbb{R}$ are necessarily identical to f.

Given that little is known about the set of maximal general dual-feasible functions, analysing the extremality of maximal dual-feasible functions becomes much more difficult, although it has been done for some functions as we show in the following example.

Example 3.4 Given the constant $c \in \{0, 1\}$, the linear function $f : \mathbb{R} \to \mathbb{R}$ with

$$f(x) := cx$$

is an extreme maximal general dual-feasible function.

This is obvious for $c = 0$, because for any general dual-feasible functions $g, h : \mathbb{R} \to \mathbb{R}$, it holds necessarily that $g(x) \leq 0$ and $h(x) \leq 0$ for $x \leq 0$. If g and h are maximal, then additionally $g(x) \geq 0$ and $h(x) \geq 0$ for all $x \geq 0$ hold, such that $f \equiv g$.

If $c = 1$, then let $g, h : \mathbb{R} \to \mathbb{R}$ be any maximal general dual-feasible functions with

$$g(x) + h(x) = 2x$$

for all $x \in \mathbb{R}$. The defining condition for dual-feasible functions implies that

$$g(1/q) \leq 1/q \quad \text{and} \quad h(1/q) \leq 1/q$$

for any $q \in \mathbb{N} \setminus \{0\}$, and hence

$$g(1/q) = 1/q.$$

The superadditivity yields

$$g(p/q) \geq p/q \quad \text{and} \quad h(p/q) \geq p/q$$

for all $p, q \in \mathbb{N} \setminus \{0\}$, and hence $g(x) = x$ for all $x \in \mathbb{Q}_+$. The monotonicity implies $g(x) = x$ for all $x \in \mathbb{R}_+$. If there would be an $x < 0$ with $g(x) > x$, then g would not be a general dual-feasible function. Therefore, $g(x) \leq x$ and finally

$$g(x) = x = f(x)$$

for all $x \in \mathbb{R}$. □

The set of maximal dual-feasible functions is convex in the bounded case of domain and range $[0, 1]$, while in the generalized case, since the symmetry needs not to hold, the set of maximal general dual-feasible functions is not convex, as counter examples show. Until now, the properties of this set have not been explored in depth, and issues as if it is at least connected, way connected or even star shaped, remain to be determined. A first result concerning the set of general dual-feasible functions is stated in the following proposition.

Proposition 3.3 *The set of general dual-feasible functions $f : \mathbb{R} \to \mathbb{R}$ is closed, i.e. any converging sequence of general dual-feasible functions converges to a general dual-feasible function.*

Proof Let I be any finite index set, and (f_n) be a converging sequence of general dual-feasible functions, i.e. for each $x \in \mathbb{R}$, the limit

$$f(x) := \lim_{n \to \infty} f_n(x)$$

exists. For any $x_i \in \mathbb{R}$ ($i \in I$) with $\sum_{i \in I} x_i \leq 1$, we have

$$\sum_{i \in I} f(x_i) = \sum_{i \in I} \lim_{n \to \infty} f_n(x_i) = \lim_{n \to \infty} \sum_{i \in I} f_n(x_i) \leq 1.$$

Therefore, f is a general dual-feasible function. □

Proposition 3.3 shows that if someone replaces a non-maximal general dual-feasible function by a dominating one and repeats this process again and again, then a general dual-feasible function will still be obtained.

3.3 Applications

In the previous chapters, we already mentioned that classical dual-feasible functions generate solutions that are feasible for the dual of the continuous relaxation of instances of the 1-dimensional cutting stock problem. General dual-feasible functions do the same for the case where the sizes of the items and the variables of the dual problem are unrestricted in sign. Negative sizes may occur, if a certain fixed quantity of extra space in containers, which can be seen as items of negative size, may be used in limited number.

Balance constraints can also lead to negative item sizes. Consider for example a process assignment problem, where p processes have to be assigned to a minimum number of identical machines, which are equipped with D units of RAM and C units of CPU capacity each. Every process i has given demands c_i in CPU and d_i in RAM, respectively. After the assignment, the processors should be well-balanced, i.e. CPU and RAM should be approximately equally exhausted, otherwise new processes could not be assigned to the machine, such that the unused resource would be wasted.

3.3 Applications

Column vectors $\mathbf{a}^j \in \mathbb{Z}_+^p$ represent configurations, i.e. processes scheduled for a single processor. The balance constraint can be written as $\mathbf{c}^\top \mathbf{a} - \mathbf{d}^\top \mathbf{a} \leq T$, where $T > 0$ is a threshold above which the machine is not considered well balanced. Setting $\mathbf{l} := \frac{1}{T}(\mathbf{c} - \mathbf{d})$, the constraint becomes $\mathbf{l}^\top \mathbf{a} \leq 1$, where the elements of $\mathbf{c} - \mathbf{d}$ can be positive or negative.

This variant can be modelled similarly to the cutting-stock problem. Let $\mathbf{b} \in \mathbb{R}_+^p$ and $\mathbf{l} \in \mathbb{R}^p$ be some fixed vectors denoting respectively the demands and the sizes of a set of p items. Each configuration may be represented by a column vector $\mathbf{a}^j \in \mathbb{Z}_+^p$ and is feasible only if it obeys the capacity constraint $\mathbf{l}^\top \mathbf{a}^j \leq 1$. The patterns form the matrix $\mathbf{A}$. The continuous relaxation of this variant of the cutting stock problem is given by

$$\min \sum_{j=1}^{n} x_j \text{ s.to } \mathbf{A}\mathbf{x} = \mathbf{b}, \ \mathbf{x} \in \mathbb{Z}_+^n. \tag{3.5}$$

If one item has a negative size, overproduction would make a trivial solution with only one bin possible. Therefore, the demand constraints must be satisfied as equalities, and the dual of (3.5) has variables unrestricted in sign

$$\max \mathbf{b}^\top \mathbf{u} \text{ s.to } \mathbf{u} \in \mathbb{R}^p, \mathbf{u}^\top \mathbf{a}^j \leq 1, \ j = 1, \ldots, n,$$

whose solutions can be obtained through general dual-feasible functions $f : \mathbb{R} \to \mathbb{R}$ by

$$u_i := f(\ell_i), \ i = 1, \ldots, p.$$

Example 3.5 Suppose that $C = D = 21$, and that seven processes with the following resource consumption have to be scheduled:

c_i	13	9	7	6	5	4	3
d_i	1	5	7	8	9	10	11

Let $T := 4$. Besides the constraints that the processes do not overload the machines, the balance constraint yields

$$(12\ 4\ 0\ -2\ -4\ -6\ -8)\mathbf{a} \leq 4.$$

Therefore, the first process must be combined with the last one, while the single balance constraint does not restrict the other processes. One may get a solution with four machines as illustrated in Fig. 3.3. The horizontal direction shows the CPU usage, the vertical direction the memory consumption on the machines. Here, one has a more-dimensional vector packing problem. In the continuous relaxation, one could use the following patterns in the quantity $1/2$:

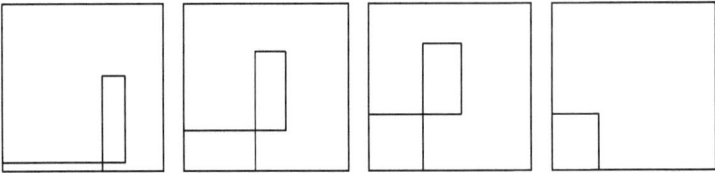

Fig. 3.3 Vector packing problem with a balance constraint

$$\mathbf{a}^1 = (1,0,0,0,0,2,0)^\top$$
$$\mathbf{a}^2 = (1,0,0,0,1,0,1)^\top$$
$$\mathbf{a}^3 = (0,1,0,2,0,0,0)^\top$$
$$\mathbf{a}^4 = (0,1,1,0,1,0,0)^\top$$
$$\mathbf{a}^5 = (0,0,1,0,0,0,1)^\top$$

where only the last pattern does not occupy the whole machine. Therefore, the continuous relaxation would yield a lower bound of two and a half needed machines. For this problem, lower bounds may be obtained by general dual-feasible functions. □

3.4 Properties of Maximal General Dual-Feasible Functions

In this section, we explore the properties of maximal general dual-feasible functions. This analysis implies frequently to prove superadditivity. Since these proofs are usually not straightforward, we will resort to intermediary results to simplify them. These results are recalled below in the form of lemmas. The first one is an extension of Lemma 2.2. The second relies on the fact that showing nonnegativity for nonnegative arguments is usually simpler than proving monotonicity.

Lemma 3.2 *Let* $f : \mathbb{R} \to \mathbb{R}$ *and* $b \in \mathbb{R}$ *with* $b > 0$ *and* $f(0) = 0$ *be given. If* f *is convex on* $[0, b]$, *then the superadditivity (3.4) holds for all* $x, y \in [0, b]$ *with* $x + y \leq b$. *If* f *is convex on* $[-b, 0]$, *then the superadditivity (3.4) holds for all* $x, y \in [-b, 0]$ *with* $x + y \geq -b$.

This lemma remains valid for $b \to \infty$, i.e. if f is convex on $\mathbb{R}_+$ and $f(0) = 0$, then f is superadditive for all nonnegative arguments. Nevertheless, such a function will usually not be a maximal general dual-feasible function. Figure 3.4 shows two real functions f, g, which are strict convex on $[-b, 0]$ and $[0, b]$, respectively. Since $f(0) = g(0) = 0$, both functions are superadditive in the specified intervals.

3.4 Properties of Maximal General Dual-Feasible Functions

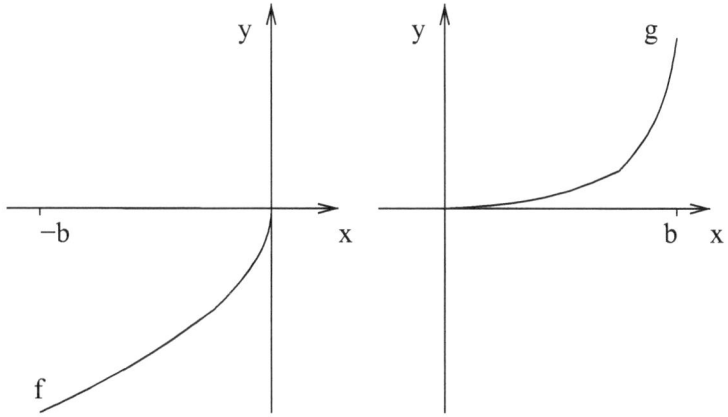

Fig. 3.4 Strict convex functions

3.4.1 Structure

A first impression about the structure of maximal general dual-feasible functions is given in the following. One will see that these functions are affine-linearly bounded. The next proposition is motivated by the following functions $f_1, f_2, f_3 : \mathbb{R} \to \mathbb{R}$, which are not maximal general dual-feasible functions:

$$f_1(x) := \begin{cases} -x^2, & \text{if } x < 0, \\ x, & \text{if } 0 \leq x \leq 1, \\ x^2, & \text{if } x > 1, \end{cases} \qquad f_2(x) := \begin{cases} -\sqrt{-x}, & \text{if } x < 0, \\ x, & \text{if } 0 \leq x \leq 1, \\ \sqrt{x}, & \text{if } x > 1, \end{cases}$$

$$f_3(x) := \begin{cases} (1 + \ln 2) \times x, & \text{if } x < 0, \\ (1 + \ln 2) \times x - \ln(1 + x), & \text{otherwise.} \end{cases}$$

The function f_1 violates the superadditivity condition, because

$$f_1(-2) = -4 < -2 = 2 \times f_1(-1),$$

while f_2 is obviously not a general dual-feasible function because of

$$5 \times 1 + (-4) \leq 1,$$

such that the defining condition on general dual-feasible functions can be applied, but

$$5 \times f_2(1) + f_2(-4) = 5 - 2 > 1.$$

Regarding f_3, one has $f_3(0) = 0$ and $f_3(1) = 1$. Moreover,

$$f_3'(x) = 1 + \ln 2 - \frac{1}{x+1} > 0,$$

for $x > 0$, such that f_3 is nondecreasing. The monotonicity of f_3' shows that f_3 is strictly convex for positive arguments, and therefore superadditive in $\mathbb{R}_+$. It is easy to see that the superadditivity of f_3 holds on entire $\mathbb{R}$. Therefore, f_3 is a general dual-feasible function fulfilling all the necessary conditions of Theorem 3.1, but it is not maximal as the next proposition states.

Proposition 3.4 *Let $f : \mathbb{R} \to \mathbb{R}$ be a maximal general dual-feasible function and*

$$t := \sup\{f(x)/x : x > 0\}.$$

Then, we have

$$\lim_{x \to \infty} \frac{f(x)}{x} = t \leq -f(-1),$$

and for any $x \in \mathbb{R}$, it holds that

$$tx - \max\{0, t-1\} \leq f(x) \leq tx,$$

i.e. f is the sum of a linear and a bounded function.

Example 3.6 The function $f_{BJ,1}$ considered in Example 3.1 yields $t = \frac{C}{\lfloor C \rfloor}$ and

$$tx - f_{BJ,1}(x) \leq t \times \frac{\text{frac}(C)}{C} = \frac{\text{frac}(C)}{\lfloor C \rfloor}.$$

Here, the mapping $x \mapsto tx - f_{BJ,1}(x)$ is a periodic, piecewise linear function, as illustrated in Fig. 3.5. □

The following lemma can be used to prove that a symmetric function f, bounded by a maximal general dual-feasible function g for arguments less than $1/2$, is a maximal general dual-feasible function if it further satisfies some additional conditions.

Lemma 3.3 *Let $f, g : \mathbb{R} \to \mathbb{R}$ be two functions with $f(1) = 1$, and both fulfilling the conditions (3.) and (4.) of part (a) of Theorem 3.1 (p. 55). If g is a maximal general dual-feasible function, and if*

$$f(x) \leq g(x) \quad \text{for all} \quad x < 1/2,$$

3.4 Properties of Maximal General Dual-Feasible Functions

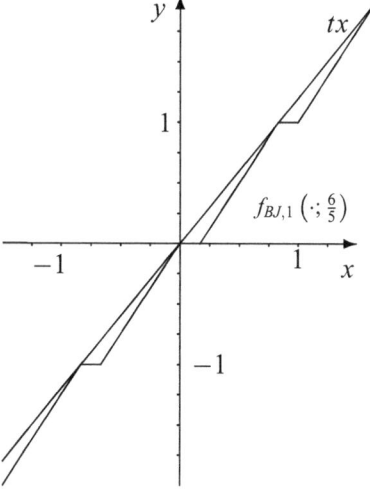

Fig. 3.5 $f_{BJ,1}$ compared to its asymptote

and

$$f(x+y) \geq f(x) + f(y) \quad \text{for all} \quad x, y \in \mathbb{R} \quad \text{with} \quad x+y < 1/2, \tag{3.6}$$

then f is a maximal general dual-feasible function.

For classical symmetric dual-feasible functions g, it was enough to have $f(x) \leq g(x)$, for a given function $f : [0, 1] \to [0, 1]$ and for all $x < 1/2$, to conclude that f is a maximal dual-feasible function. For general dual-feasible functions, the situation changes. Nevertheless, Lemma 3.3 gives the possibility to demand the superadditivity of the symmetric function f only for $x + y < 1/2$, instead of all x, y with

$$x \leq y \leq \frac{1-x}{2},$$

as it would be necessary according to part (d) of Theorem 3.1. The following example shows that the prerequisite on superadditivity must not be dropped entirely.

Example 3.7 Let g be the identity function and

$$f(x) := \begin{cases} x^3, & \text{if } x \leq -1, \\ x, & \text{if } -1 \leq x \leq 2, \\ (x-1)^3 + 1, & \text{if } x \geq 2. \end{cases}$$

The functions f and g fulfill all prerequisites except the superadditivity condition. One has $f(1) = 1$, $f(x) \leq g(x)$ for all $x \leq 2$, both f and g are symmetric, and g is a maximal general dual-feasible function. However, f violates the last prerequisite of Lemma 3.3. Since f is not a general dual-feasible function, this example shows the need of the restricted superadditivity (3.6). □

3.4.2 Behaviour at Given Points

Every classical maximal dual-feasible function f is symmetric, and therefore it fulfills the conditions

$$f(1) = 1 \text{ and } f(1/2) = 1/2.$$

Even if former proofs for the latter condition cannot be used for maximal general dual-feasible functions, the two conditions hold for many maximal general dual-feasible functions. Furthermore, they are not only necessary conditions, but they allow also immediately the lower bound

$$f(x) \geq \lfloor 2x \rfloor / 2$$

for $x > 1$ for these functions due to the superadditivity. An insight concerning the reasons why not all approaches for classical maximal dual-feasible functions work is given in Exercise 3.

Proposition 3.5 *If $f : \mathbb{R} \to \mathbb{R}$ is a maximal general dual-feasible function and not of the kind $x \mapsto tx$ with $0 \leq t < 1$, then $f(1) = 1$ and $f(1/2) = 1/2$.*

Proof Let $t := \sup\{f(x)/x : x > 0\} \geq 0$. If $t < 1$, then Proposition 3.4 implies $f(x) = tx$ for all $x \in \mathbb{R}$, such that the proof is complete. Otherwise $t \geq 1$ and $f(x) \geq tx + 1 - t$ for all $x \in \mathbb{R}$, and $f(1) \geq 1$. Therefore, we have $f(1) = 1$. Since f is a maximal general dual-feasible function, it is nondecreasing and superadditive. Suppose $f(1/2) < 1/2$ (clearly, $f(1/2) > 1/2$ is impossible). Define $g : \mathbb{R} \to \mathbb{R}$ by

$$g(x) := \begin{cases} f(x), & \text{if } x \neq 1/2, \\ 1/2, & \text{otherwise.} \end{cases}$$

Due to the assumption, g cannot be a general dual-feasible function, i.e. there are values $n \in \mathbb{N} \setminus \{0\}$ and $x_1, \ldots, x_n \in \mathbb{R}$ with $\sum_{i=1}^{n} x_i \leq 1$, but $\sum_{i=1}^{n} g(x_i) > 1$. Without loss of generality, assume $x_i = 1/2$ for $i \leq k$ and $x_i \neq 1/2$ for $i > k$ with a certain $k \in \mathbb{N}, k \leq n$. That yields

$$\sum_{i=k+1}^{n} x_i \leq 1 - k/2$$

3.4 Properties of Maximal General Dual-Feasible Functions

and

$$1 < \sum_{i=1}^{n} g(x_i) = k/2 + \sum_{i=k+1}^{n} f(x_i)$$
$$\leq k/2 + f\left(\sum_{i=k+1}^{n} x_i\right)$$
$$\leq k/2 + f(1 - k/2),$$

where empty sums equal zero. Since $f(1) = 1$, one gets further

$$1 < \operatorname{frac}(k/2) + \lfloor k/2 \rfloor \times 1 + f(1 - k/2) \leq \operatorname{frac}(k/2) + f(1 - k/2 + \lfloor k/2 \rfloor \times 1)$$
$$= \operatorname{frac}(k/2) + f(1 - \operatorname{frac}(k/2)).$$

If k is even, then the contradiction $1 < 0 + f(1)$ arises. If k is odd, then

$$1 < 1/2 + f(1/2)$$

yields also a contradiction. □

The following proposition characterises the behaviour of a maximal general dual-feasible function at the point zero, and it can help to prove or disprove that a given real function is a maximal general dual-feasible function.

Proposition 3.6 *If $f : \mathbb{R} \to \mathbb{R}$ is a maximal general dual-feasible function, then, for any $y \in \mathbb{R}$, the left and right limits $\lim_{x \uparrow y} f(x)$ and $\lim_{x \downarrow y} f(x)$ exist, and it holds that*

$$\lim_{x \uparrow 0} f(x) = \inf_{y \in \mathbb{R}} \left\{ \lim_{x \uparrow y} f(x) - \lim_{x \downarrow y} f(x) \right\}.$$

The idea is that if f has a gap at the point y, then the superadditivity requires for negative $x \approx 0$ that $f(x)$ is small enough. Meanwhile, $f(x)$ must also not be too small, otherwise f will not be maximal.

Example 3.8 Define the function $f : \mathbb{R} \to \mathbb{R}$ by

$$f(x) := \begin{cases} b \times \lfloor 2x \rfloor, & \text{if } x < 1/2, \\ 1/2, & \text{if } x = 1/2, \\ 1 - b \times \lfloor 2 - 2x \rfloor, & \text{if } x > 1/2, \end{cases} \quad (3.7)$$

which is a maximal general dual-feasible function for every $b \geq 1$, as can be proved by Theorem 3.1. This discontinuous function has as its largest gap b, namely at points y, where $2y \in \mathbb{Z}$. That fits to

$$\lim_{x \uparrow 0} f(x) = b \times (-1) = -b.$$

□

3.4.3 Limits of Possible Convexity

The following proposition demonstrates the limits of possible convexity for maximal general dual-feasible functions.

Proposition 3.7 *If $f : \mathbb{R} \to \mathbb{R}$ is a maximal general dual-feasible function, then it cannot be strict convex in an environment of the point zero and also not concave (except linear) in an interval $[0, b]$ or $[-b, 0]$ with $b > 0$.*

Example 3.9 Let $b := 1/2$ and

$$f(x) := 4x^3$$

for $x \in [-b, b]$. Then, f is strictly concave in the interval $[-b, 0]$, and hence it is not superadditive. For instance, we have that

$$f(-1/4) = -1/16,$$

and also

$$f(-1/2) = -1/2 < 2 \times f(-1/4).$$

□

3.4.4 Composition and Convex Combinations

While the composition of general dual-feasible functions yields a general dual-feasible function, maximality may be lost in this process as shown in the following proposition. The same happens with the convex combination of functions.

Proposition 3.8 *The composition or convex combination of maximal general dual-feasible functions is not necessarily a maximal general dual-feasible function.*

Proof Let $b, c \in \mathbb{R}$ be any constants with $0 < c < 1$ and $bc \geq 1$. The function $f : \mathbb{R} \to \mathbb{R}$ defined as $f(x) := cx$ is a maximal general dual-feasible function (see Proposition 3.2). Let g be the function (3.7). That yields the composition $f(g(\cdot)) = c \times g(\cdot)$, i.e.

$$f(g(x)) = \begin{cases} bc \times \lfloor 2x \rfloor, & \text{if } x < 1/2, \\ c/2, & \text{if } x = 1/2, \\ c - bc \times \lfloor 2 - 2x \rfloor, & \text{if } x > 1/2, \end{cases}$$

3.5 Examples

$$\leq \begin{cases} bc \times \lfloor 2x \rfloor, & \text{if } x < 1/2, \\ 1/2, & \text{if } x = 1/2, \\ 1 - bc \times \lfloor 2 - 2x \rfloor, & \text{if } x > 1/2, \end{cases}$$

and for $x \geq 1/2$ the inequality is strict. The latter function is of the same type as the function (3.7) with parameter $bc \geq 1$, and hence it is a maximal general dual-feasible function. The function $f(g(\cdot))$ can also be seen as a convex combination of the constant zero-function $f \equiv 0$, which is a maximal general dual-feasible function according to Proposition 3.2, with factor $1 - c$ and g with factor c. □

3.5 Examples

As already mentioned, many classical dual-feasible functions cannot be extended easily to general dual-feasible functions or at least not with the same formula. Such an example is given next.

Example 3.10 The function $f_{CCM,1}(\cdot; C)$ already described in Chap. 2 (see p. 28) is defined for any real parameter $C \geq 1$ as follows:

$$f_{CCM,1}(x; C) = \begin{cases} \lfloor Cx \rfloor / \lfloor C \rfloor, & \text{if } x < 1/2, \\ 1/2, & \text{if } x = 1/2, \\ 1 - \lfloor C - Cx \rfloor / \lfloor C \rfloor, & \text{if } x > 1/2. \end{cases}$$

This function was a maximal dual-feasible function for the domain $[0, 1]$. However, for domain $\mathbb{R}$, it is not even a general dual-feasible function. To see this, let

$$n := \left\lceil \frac{1}{1 - \text{frac}(C)} \right\rceil + 1$$

and choose an enough small $\varepsilon > 0$. One gets

$$f_{CCM,1}(1 + \varepsilon) = 1 + \frac{1}{\lfloor C \rfloor}$$

and

$$f_{CCM,1}\left(\frac{1 - n - n \times \lfloor C \rfloor}{C}\right) = \frac{1 - n}{\lfloor C \rfloor} - n.$$

Hence,

$$n \times f_{CCM,1}(1 + \varepsilon) + f_{CCM,1}\left(\frac{1 - n - n \times \lfloor C \rfloor}{C}\right) = \frac{1}{\lfloor C \rfloor} > 0.$$

Since

$$n \times (1 + \varepsilon) + \frac{1 - n - n \times \lfloor C \rfloor}{C} = n\varepsilon + \frac{1}{C} \times (Cn + 1 - n - n \times \lfloor C \rfloor)$$
$$= n\varepsilon + \frac{1 - n \times (1 - \text{frac}(C))}{C}$$
$$\leq n\varepsilon + \frac{1 - 1 + \text{frac}(C) - 1}{C}$$
$$< 0$$

for sufficiently small ε, a contradiction to the point (2.) of Proposition 3.1 arises. $\square$

On the other hand, we explore in the sequel a different case where the defining formulation of a classical maximal dual-feasible function leads without any change to a maximal general dual-feasible function. That happens in particular with the function $f_{DG,1}$ discussed in Chap. 2 (see (2.14) at page 43).

Proposition 3.9 *For any $C \in \mathbb{R} \setminus \mathbb{N}$, $C > 1$ and $k \in \mathbb{N}$ with $k \geq \frac{1}{\text{frac}(C)}$, the following function $f_{DG,1} : \mathbb{R} \to \mathbb{R}$ is a maximal general dual-feasible function:*

$$f_{DG,1}(x) := \frac{\lfloor Cx \rfloor}{\lfloor C \rfloor} + \frac{1}{\lfloor C \rfloor} \times \begin{cases} \frac{\text{frac}(Cx) - \text{frac}(C)}{1 - \text{frac}(C)}, & \text{if } (k-1) \times \frac{\text{frac}(Cx) - \text{frac}(C)}{1 - \text{frac}(C)} \in \mathbb{N}, \\ \max\left\{0, \lceil (k-1) \times \frac{\text{frac}(Cx) - \text{frac}(C)}{1 - \text{frac}(C)} \rceil / k\right\}, & \text{otherwise.} \end{cases}$$

Proof Define the auxiliary function $h : \mathbb{R} \to \mathbb{R}$ as

$$h(x) := \begin{cases} x, & \text{if } (k-1) \times x \in \mathbb{N}, \\ \max\{0, \lceil (k-1) \times x \rceil / k\}, & \text{otherwise.} \end{cases} \quad (3.8)$$

Then, we have

$$f_{DG,1}(x) = \frac{\lfloor Cx \rfloor + h\left(\frac{\text{frac}(Cx) - \text{frac}(C)}{1 - \text{frac}(C)}\right)}{\lfloor C \rfloor}.$$

First, some properties of h are derived, which will be used later to prove the sufficient conditions of Theorem 3.1. Clearly, one obtains $h(x) = 0$ for $x \leq 0$ and $h(1) = 1$. Moreover, h rises monotonely in the closed interval $[0, 1]$, because h is piecewise constant and one gets for any $p \in \{1, \ldots, k-1\}$ the estimations

$$\lim_{x \uparrow \frac{p}{k-1}} h(x) = \frac{p}{k} < \frac{p}{k-1} = h\left(\frac{p}{k-1}\right) \leq \frac{p+1}{k} = \lim_{x \downarrow \frac{p}{k-1}} h(x).$$

3.5 Examples

Additionally, we have

$$\frac{\lceil (k-1)x \rceil}{k} \le h(x) \le \frac{\lfloor (k-1)x \rfloor + 1}{k}, \quad \text{for all} \quad x \in [0,1], \tag{3.9}$$

due to the definition of h, because either $h(x) = \lceil (k-1)x \rceil /k$ or $h(x) = x$. In the latter case, it holds that $\lceil (k-1)x \rceil = (k-1)x$, such that both inequalities become equivalent to $x \ge 0$ or $x \le 1$, respectively. The function h is also symmetric inside the interval $[0,1]$, i.e.

$$h(x) + h(1-x) = 1, \quad \text{for all} \quad x \in [0,1], \tag{3.10}$$

because

$$(k-1) \times x \in \mathbb{N} \iff (k-1) \times (1-x) \in \mathbb{N},$$

and if $(k-1) \times x \notin \mathbb{N}$, then

$$\begin{aligned} h(x) + h(1-x) &= \lceil (k-1) \times x \rceil /k + \lceil (k-1) \times (1-x) \rceil /k \\ &= \lceil (k-1) \times x + (k-1) \times (1-x) + 1 \rceil /k \\ &= 1. \end{aligned}$$

Clearly, the function $f_{DG,1}$ fulfills the conditions (1.) and (3.) of part (a) of Theorem 3.1. To show the symmetry, choose any $x \in \mathbb{R}$. It must be verified that

$$\lfloor Cx \rfloor + \lfloor C-Cx \rfloor + h\left(\frac{\mathrm{frac}(Cx) - \mathrm{frac}(C)}{1 - \mathrm{frac}(C)} \right) + h\left(\frac{\mathrm{frac}(C-Cx) - \mathrm{frac}(C)}{1 - \mathrm{frac}(C)} \right) = \lfloor C \rfloor.$$

This is obvious for $\mathrm{frac}(Cx) = \mathrm{frac}(C)$. If $\mathrm{frac}(Cx) < \mathrm{frac}(C)$, then

$$0 < \mathrm{frac}(C - Cx) \le \mathrm{frac}(C),$$

and hence

$$\begin{aligned} f_{DG,1}(x) + f_{DG,1}(1-x) &= \frac{\lfloor Cx \rfloor + \lfloor C - Cx \rfloor}{\lfloor C \rfloor} \\ &= 1. \end{aligned}$$

If $\mathrm{frac}(Cx) > \mathrm{frac}(C)$, then

$$\lfloor Cx \rfloor + \lfloor C - Cx \rfloor = \lfloor C \rfloor - 1$$

and

$$h\left(\frac{\operatorname{frac}(C-Cx)-\operatorname{frac}(C)}{1-\operatorname{frac}(C)}\right) = h\left(\frac{\operatorname{frac}(C)-\operatorname{frac}(Cx)+1-\operatorname{frac}(C)}{1-\operatorname{frac}(C)}\right)$$
$$= h\left(1-\frac{\operatorname{frac}(Cx)-\operatorname{frac}(C)}{1-\operatorname{frac}(C)}\right)$$
$$= 1-h\left(\frac{\operatorname{frac}(Cx)-\operatorname{frac}(C)}{1-\operatorname{frac}(C)}\right),$$

such that $f_{DG,1}$ is symmetric also in this case.

Showing the superadditivity requires for any $x, y \in \mathbb{R}$ to verify that

$$d := \lfloor C \rfloor \times (f_{DG,1}(x+y) - f_{DG,1}(x) - f_{DG,1}(y))$$

is not negative. For this purpose, let

$$I_1 := \left\{ x \in \mathbb{R} : (k-1) \times \frac{\operatorname{frac}(Cx)-\operatorname{frac}(C)}{1-\operatorname{frac}(C)} \in \mathbb{N} \right\}.$$

Without loss of generality, suppose that $\operatorname{frac}(Cx) \leq \operatorname{frac}(Cy)$. Four cases arise:

1. If $\operatorname{frac}(Cy) \leq \operatorname{frac}(C)$, then

$$d \geq \lfloor Cx+Cy \rfloor - \lfloor Cx \rfloor - \lfloor Cy \rfloor$$
$$\geq 0,$$

because the rounding-down is superadditive.

2. The case where $\operatorname{frac}(Cx) \leq \operatorname{frac}(C) < \operatorname{frac}(Cy)$ yields two subcases. If $\operatorname{frac}(Cx) + \operatorname{frac}(Cy) < 1$, then

$$d = h(\frac{\operatorname{frac}(Cx+Cy)-\operatorname{frac}(C)}{1-\operatorname{frac}(C)}) - h(\frac{\operatorname{frac}(Cy)-\operatorname{frac}(C)}{1-\operatorname{frac}(C)})$$
$$\geq 0,$$

because of the monotonicity of h in the interval $(0, 1)$. If $\operatorname{frac}(Cx)+\operatorname{frac}(Cy) \geq 1$, then

$$d \geq 1 - h\left(\frac{\operatorname{frac}(Cy)-\operatorname{frac}(C)}{1-\operatorname{frac}(C)}\right)$$
$$> 0.$$

3. The case $\operatorname{frac}(Cx) > \operatorname{frac}(C)$ and $\operatorname{frac}(Cx) + \operatorname{frac}(Cy) < 1$ can only happen if $\operatorname{frac}(C) < 1/2$. One gets

$$\lfloor Cx+Cy \rfloor = \lfloor Cx \rfloor + \lfloor Cy \rfloor,$$

3.5 Examples

and hence

$$d = h\left(\frac{\text{frac}(Cx) + \text{frac}(Cy) - \text{frac}(C)}{1 - \text{frac}(C)}\right) - h\left(\frac{\text{frac}(Cx) - \text{frac}(C)}{1 - \text{frac}(C)}\right)$$
$$- h\left(\frac{\text{frac}(Cy) - \text{frac}(C)}{1 - \text{frac}(C)}\right).$$

The estimation (3.9) yields

$$dk \geq \left\lceil (k-1) \times \frac{\text{frac}(Cx) + \text{frac}(Cy) - \text{frac}(C)}{1 - \text{frac}(C)}\right\rceil$$
$$- \left\lfloor (k-1) \times \frac{\text{frac}(Cx) - \text{frac}(C)}{1 - \text{frac}(C)}\right\rfloor$$
$$- 1 - \left\lfloor (k-1) \times \frac{\text{frac}(Cy) - \text{frac}(C)}{1 - \text{frac}(C)}\right\rfloor - 1 \quad (3.11)$$

$$\geq \frac{k-1}{1 - \text{frac}(C)} \times (\text{frac}(Cx) + \text{frac}(Cy) - \text{frac}(C) - \text{frac}(Cx) + \text{frac}(C)$$
$$- \text{frac}(Cy) + \text{frac}(C)) - 2$$

$$= \frac{k \times \text{frac}(C) - \text{frac}(C)}{1 - \text{frac}(C)} - 2$$

$$\geq \frac{1 - \text{frac}(C)}{1 - \text{frac}(C)} - 2 \quad \text{(because of } k \geq \tfrac{1}{\text{frac}(C)}\text{)}$$

$$= -1.$$

The right-hand side of (3.11) is integer. It may be equal to -1 only, if the rounding-brackets had no influence. If $x \notin I_1$ or $y \notin I_1$, then the rounding changed the values, such that the right-hand side of (3.11) is at least zero, and hence $d \geq 0$. If $x, y \in I_1$, then

$$h\left(\frac{\text{frac}(Cx) - \text{frac}(C)}{1 - \text{frac}(C)}\right) + h\left(\frac{\text{frac}(Cy) - \text{frac}(C)}{1 - \text{frac}(C)}\right)$$
$$= \frac{\text{frac}(Cx) + \text{frac}(Cy) - 2\text{frac}(C)}{1 - \text{frac}(C)}$$
$$= h\left(\frac{\text{frac}(Cx + Cy) - 2\text{frac}(C)}{1 - \text{frac}(C)}\right)$$
$$\leq h\left(\frac{\text{frac}(Cx + Cy) - \text{frac}(C)}{1 - \text{frac}(C)}\right)$$

due to the monotonicity of h inside the interval $[0, 1]$, and hence, we have again $d \geq 0$.

4. When $\text{frac}(Cx) > \text{frac}(C)$ and $\text{frac}(Cx) + \text{frac}(Cy) \geq 1$, we get

$$d = \lfloor Cx + Cy \rfloor + h\left(\frac{\text{frac}(Cx + Cy) - \text{frac}(C)}{1 - \text{frac}(C)}\right) - \lfloor Cx \rfloor - h\left(\frac{\text{frac}(Cx) - \text{frac}(C)}{1 - \text{frac}(C)}\right)$$
$$- \lfloor Cy \rfloor - h\left(\frac{\text{frac}(Cy) - \text{frac}(C)}{1 - \text{frac}(C)}\right)$$
$$= 1 + h\left(\frac{\text{frac}(Cx) + \text{frac}(Cy) - 1 - \text{frac}(C)}{1 - \text{frac}(C)}\right) - h\left(\frac{\text{frac}(Cx) - \text{frac}(C)}{1 - \text{frac}(C)}\right)$$
$$- h\left(\frac{\text{frac}(Cy) - \text{frac}(C)}{1 - \text{frac}(C)}\right).$$

If $\frac{\text{frac}(Cx) - \text{frac}(C)}{1 - \text{frac}(C)} + \frac{\text{frac}(Cy) - \text{frac}(C)}{1 - \text{frac}(C)} \leq 1$, then the monotonicity of h inside the interval $[0, 1]$ and (3.10) yield

$$h\left(\frac{\text{frac}(Cx) - \text{frac}(C)}{1 - \text{frac}(C)}\right) + h\left(\frac{\text{frac}(Cy) - \text{frac}(C)}{1 - \text{frac}(C)}\right) \leq 1,$$

such that

$$d \geq h\left(\frac{\text{frac}(Cx) + \text{frac}(Cy) - 1 - \text{frac}(C)}{1 - \text{frac}(C)}\right)$$
$$\geq 0.$$

Therefore assume for the rest of the proof that

$$\frac{\text{frac}(Cx) - \text{frac}(C)}{1 - \text{frac}(C)} + \frac{\text{frac}(Cy) - \text{frac}(C)}{1 - \text{frac}(C)} > 1.$$

One has

$$d = 1 + h\left(\frac{\text{frac}(Cx) + \text{frac}(Cy) - 2\text{frac}(C)}{1 - \text{frac}(C)} - 1\right) - h\left(\frac{\text{frac}(Cx) - \text{frac}(C)}{1 - \text{frac}(C)}\right)$$
$$- h\left(\frac{\text{frac}(Cy) - \text{frac}(C)}{1 - \text{frac}(C)}\right).$$

Analogously to the previous case, (3.9) yields

$$dk \geq k + \left\lceil (k-1) \times \frac{\text{frac}(Cx) + \text{frac}(Cy) - 2\text{frac}(C)}{1 - \text{frac}(C)} - k + 1 \right\rceil$$
$$- \left\lfloor (k-1) \times \frac{\text{frac}(Cx) - \text{frac}(C)}{1 - \text{frac}(C)} \right\rfloor - 1 - \left\lfloor (k-1) \times \frac{\text{frac}(Cy) - \text{frac}(C)}{1 - \text{frac}(C)} \right\rfloor - 1$$

(3.12)

3.6 Building Maximal General Dual-Feasible Functions

$$\geq \frac{k-1}{1-\operatorname{frac}(C)} \times (\operatorname{frac}(Cx) + \operatorname{frac}(Cy) - 2\operatorname{frac}(C) - \operatorname{frac}(Cx) + \operatorname{frac}(C)$$
$$- \operatorname{frac}(Cy) + \operatorname{frac}(C)) - 1$$
$$= -1,$$

and the right-hand side in (3.12) is integer. If it equals -1, then the rounding brackets will not change anything, and hence $x, y \in I_1$. In this case, one gets

$$h\left(\frac{\operatorname{frac}(Cx) - \operatorname{frac}(C)}{1 - \operatorname{frac}(C)}\right) +$$
$$+ h\left(\frac{\operatorname{frac}(Cy) - \operatorname{frac}(C)}{1 - \operatorname{frac}(C)}\right) - 1 = \frac{\operatorname{frac}(Cx) - \operatorname{frac}(C)}{1 - \operatorname{frac}(C)} + \frac{\operatorname{frac}(Cy) - \operatorname{frac}(C)}{1 - \operatorname{frac}(C)} - 1$$
$$= \frac{\operatorname{frac}(Cx) + \operatorname{frac}(Cy) - 2\operatorname{frac}(C)}{1 - \operatorname{frac}(C)} - 1$$
$$= h\left(\frac{\operatorname{frac}(Cx) + \operatorname{frac}(Cy) - 2\operatorname{frac}(C)}{1 - \operatorname{frac}(C)} - 1\right),$$

such that again $d \geq 0$.

Since in all cases it holds that $d \geq 0$, the proof is complete. □

3.6 Building Maximal General Dual-Feasible Functions

In this section, we explore three different methods to build maximal general dual-feasible functions by extending a given classical dual-feasible function to domain and range $\mathbb{R}$. The second and third approach simply use affine-linear expressions outside the interval $[0, 1]$ where the extended function has the form

$$x \mapsto tx + c,$$

with t and c being constants, and t a sufficiently large value. The additive constants are generally different for $x < 0$ and $x > 1$.

3.6.1 Method I

Classical maximal dual-feasible functions can be extended to domain and range $\mathbb{R}$ by considering the sum of a periodic with a monotone staircase function whose average slope must be large enough. The feasible lower bound of this average slope can be expressed in different ways. The classical maximal dual-feasible function is applied to the non-integer part of the given real argument, while a suitable multiple of the integer part of the argument is added. These ideas are stated formally in the following proposition.

Proposition 3.10 *Let $g : [0, 1] \to [0, 1]$ be a maximal dual-feasible function,*

$$h(x, y) := g(x + y) - g(x) - g(y) \text{ with } x, y \in [0, 1] \text{ and } x + y \leq 1, \quad (3.13)$$

and

$$b_0 := \sup\{g(x) + g(y) - g(x + y - 1) \mid x, y \in [0, 1] \wedge x + y \geq 1\}.$$

Then, we have that

$$b_0 = 1 + \sup\{h(x, y) : 0 < x \leq y < 1/2 \text{ and } x + y \leq 2/3\}$$

and

$$1 \leq b_0 \leq 2.$$

For $b \geq b_0$, g can be extended to a maximal general dual-feasible function $f : \mathbb{R} \to \mathbb{R}$ as follows

$$f(x) := \begin{cases} g(\mathsf{frac}(x)) + b \times \lfloor x \rfloor, & \text{if } x < 1, \\ 1 - f(1 - x), & \text{otherwise.} \end{cases}$$

For $b > b_0$, f becomes a non-trivial convex combination of maximal general dual-feasible functions. In that case, f is clearly not extreme.

To see this, choose any $\lambda \in (0, b - b_0]$ and set $b_1 := b - \lambda$ and $b_2 := b + \lambda$. We may use the proposition with b_1 and b_2 to get the non-identical functions f, f_1, f_2. One obtains for any $x < 1$ that

$$2f(x) - f_1(x) - f_2(x) = 2 \times (g(\mathsf{frac}(x)) + b \times \lfloor x \rfloor) - g(\mathsf{frac}(x)) - b_1 \times \lfloor x \rfloor$$
$$- g(\mathsf{frac}(x)) - b_2 \times \lfloor x \rfloor$$
$$= \lfloor x \rfloor \times (2b - b_1 - b_2)$$
$$= 0.$$

If $x \geq 1$, then the symmetry leads to the same result.

Example 3.11 For any $k \in \mathbb{N} \setminus \{0\}$, the function $f_{FS,1}(\cdot; k) : [0, 1] \to [0, 1]$ (already described in Chap. 2, see (2.10), page 35) is a classical maximal dual feasible function, which is defined as follows

$$f_{FS,1}(x; k) = \begin{cases} x, & \text{if } (k + 1) \times x \in \mathbb{N}, \\ \lfloor (k + 1) \times x \rfloor / k, & \text{otherwise.} \end{cases} \quad (3.14)$$

3.6 Building Maximal General Dual-Feasible Functions

Let g be this function $f_{FS,1}$ in Proposition 3.10. Formula (3.13) yields

$$h\left(\frac{1}{k+1} - \varepsilon, \frac{1}{k+1} - \varepsilon\right) = f_{FS,1}\left(\frac{2}{k+1} - 2\varepsilon\right) - 2f_{FS,1}\left(\frac{1}{k+1} - \varepsilon\right)$$
$$= \frac{1}{k}$$

for enough small $\varepsilon > 0$. For the chosen function g, greater values of h are not possible, i.e.

$$\max\{h(x, y) \mid 0 \leq x, y \leq 1 \text{ and } x + y \leq 1\} = 1/k.$$

Therefore, one gets $b_0 = \frac{k+1}{k}$, and according to Proposition 3.10 with $b := b_0$ the function $f : \mathbb{R} \to \mathbb{R}$ with

$$f(x) := \begin{cases} \text{frac}(x) + \frac{k+1}{k} \times \lfloor x \rfloor, & \text{if } (k+1) \times \text{frac}(x) \in \mathbb{N} \land x < 1, \\ \frac{\lfloor (k+1) \times \text{frac}(x) \rfloor}{k} + \frac{k+1}{k} \times \lfloor x \rfloor, & \text{if } (k+1) \times \text{frac}(x) \notin \mathbb{N} \land x < 1, \\ 1 - f(1-x), & \text{if } x \geq 1, \end{cases}$$

is a maximal general dual-feasible function. This function is illustrated in Fig. 3.6 for $k \in \{1, 2\}$. □

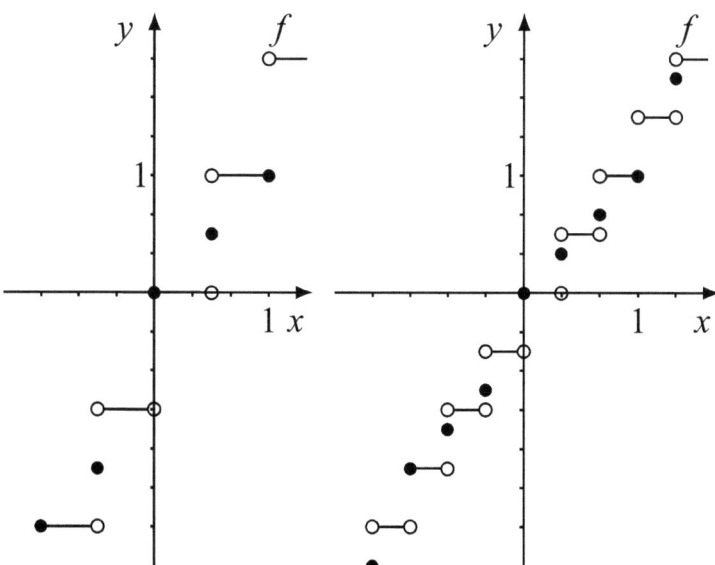

Fig. 3.6 Extending $f_{FS,1}(\cdot; k)$ to a general MDFF f for $k \in \{1, 2\}$ by Proposition 3.10

3.6.2 Method II

The next proposition allows the extension of a Lipschitz-continuous maximal dual-feasible function $g : [0, 1] \to [0, 1]$ to the domain and range $\mathbb{R}$ by using affine-linear terms for the arguments that are outside $[0, 1]$. A function $f : X \subseteq \mathbb{R} \to \mathbb{R}$ is *Lipschitz-continuous*, if there is a constant $L \geq 0$ with

$$|f(x) - f(y)| \leq L \times |x - y| \text{ for all } x, y \in X.$$

Lipschitz-continuity is a strong property that can be generalized to higher-dimensional Euclidean spaces without difficulty. Let $X \subseteq \mathbb{R}^n$ be an open set, $n \in \mathbb{N}$. According to Rademacher's theorem, every Lipschitz-continuous function $f : X \to \mathbb{R}^m$, $m \in \mathbb{N}$, is almost everywhere totally differentiable. Furthermore, a differentiable function is Lipschitz-continuous if and only if its derivative is bounded. Moreover, we have the following implications, where the converse is false:

– every Lipschitz-continuous function f is *Hölder*-continuous, i.e. there are constants $c > 0$ and $\alpha \in (0, 1]$ with

$$|f(x) - f(y)| \leq c \times |x - y|^\alpha \text{ for all } x, y;$$

– every Hölder-continuous function f is uniformly continuous, i.e., for all $\varepsilon > 0$, there is a $\delta > 0$ depending on ε only, such that $|x - y| < \delta$ implies

$$|f(x) - f(y)| < \varepsilon;$$

– every uniform continuous function is also continuous in the usual sense, where δ may depend on x and y. Here, the converse holds under an additional prerequisite. Every continuous function on a compact set is uniformly bounded according to Heine's theorem.

Examples of Lipschitz-continuous functions are $x \mapsto |x|$ and $x \mapsto e^{|x|}$ on a bounded set, while the function $x \mapsto \sqrt{\max\{0, x\}}$, for example, is not Lipschitz-continuous.

Proposition 3.11 *Let $p, t \in \mathbb{R}$ and $g : [0, 1] \to [0, 1]$ be a maximal dual-feasible function with*

$$|g(x) - g(y)| \leq t \times |x - y|$$

3.6 Building Maximal General Dual-Feasible Functions

for all $x, y \in [0, 1]$, i.e. the function g is Lipschitz-continuous with $L := t$. Then, we have $t \geq 1$, and, for $1 \leq p \leq t$, the following function $f : \mathbb{R} \to \mathbb{R}$ is a maximal general dual-feasible function:

$$f(x) := \begin{cases} tx + 1 - p, & \text{if } x < 0, \\ tx + p - t, & \text{if } x > 1, \\ g(x), & \text{otherwise.} \end{cases}$$

Exercise 8 is an example that weakening the Lipschitz-continuity to Hölder-continuity can cause the loss of superadditivity.

Example 3.12 The following function $g : [0, 1] \to [0, 1]$ is symmetric, strict convex in $[0, \frac{1}{2}]$ and a Lipschitz-continuous maximal dual-feasible function:

$$g(x) := \begin{cases} 2x^2, & \text{if } 0 \leq x \leq \frac{1}{2}, \\ 1 - g(1 - x), & \text{otherwise.} \end{cases}$$

To calculate the smallest valid Lipschitz-constant for g, the largest slope is needed. Since g is differentiable in $(0, 1)$, the supremum of the derivative is sought. One gets

$$g'(x) = 4x$$

inside the interval $(0, \frac{1}{2})$. Because of the symmetry of g, it holds that

$$g'(1 - x) = g'(x),$$

such that g' is continuous at $1/2$, and hence it is enough to analyze g (or g') for $x \leq \frac{1}{2}$ only. The derivative g' takes its maximum at $x = 1/2$ with $g'\left(\frac{1}{2}\right) = 2$. Hence, the smallest valid Lipschitz-constant is 2. The extension of g to a maximal general dual-feasible function $f : \mathbb{R} \to \mathbb{R}$ for $t = 2$ and $p \in \{1, t\}$ according to Proposition 3.11 is illustrated in Fig. 3.7.

□

3.6.3 Method III

Most of the classical maximal dual-feasible functions are not Lipschitz-continuous, such that Proposition 3.11 cannot be used in these cases. In this section, we introduce a similar proposition that does not need this strong prerequisite. Some generality is lost in the construction, because instead of two parameters only one will apply. Here again, affine-linear expressions are used outside the interval $[0, 1]$ to extend a given classical maximal dual-feasible function to a general one.

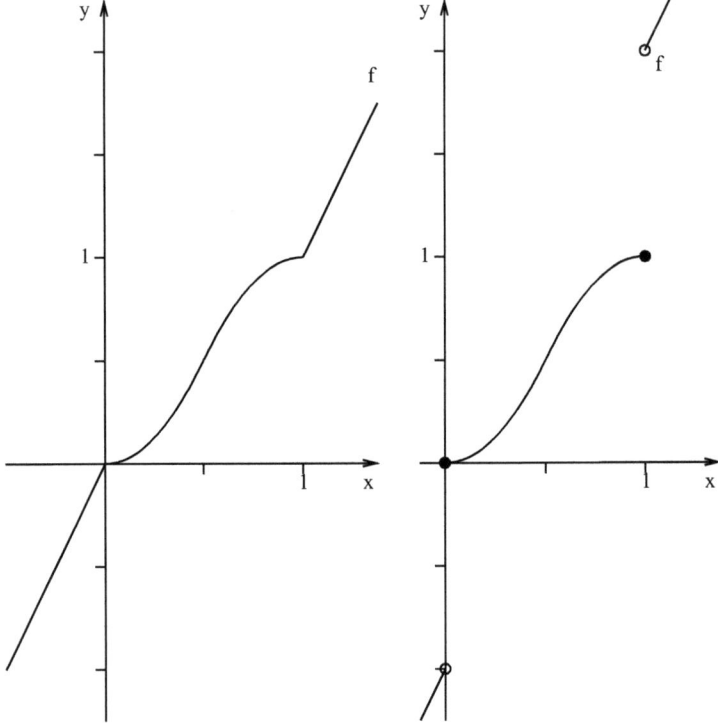

Fig. 3.7 Application of Proposition 3.11 to g with $t = 2$ and $p \in \{1, t\}$

Proposition 3.12 *Let $t \in \mathbb{R}$ and $g : [0, 1] \to [0, 1]$ be a maximal dual-feasible function with $g(x) \leq tx$ for all $x \in [0, 1]$. Then, we have $t \geq 1$, and the following function $f : \mathbb{R} \to \mathbb{R}$ is a maximal general dual-feasible function:*

$$f(x) := \begin{cases} tx + 1 - t, & \text{if } x < 0, \\ tx, & \text{if } x > 1, \\ g(x), & \text{otherwise.} \end{cases}$$

Example 3.13 Let us again use the function (3.14) as example. One gets for $p \in \mathbb{N}$ and $p \leq k$ that

$$\lim_{x \downarrow \frac{p}{k+1}} f_{FS,1}(x) = \frac{p}{k}.$$

Hence, the smallest possible t for Proposition 3.12 is

$$t = \frac{k+1}{k},$$

3.6 Building Maximal General Dual-Feasible Functions

because if $(k+1) \times x \in \mathbb{N}$ for a certain $x \in [0,1]$, then $f_{FS,1}(x) = x$. With these choices, one obtains the following maximal general dual-feasible function $f : \mathbb{R} \to \mathbb{R}$:

$$f(x) := \begin{cases} tx + 1 - t, & \text{if } x < 0, \\ tx, & \text{if } x > 1, \\ x, & \text{if } (k+1) \times x \in \{0, 1, \ldots, k+1\}, \\ \lfloor (k+1) \times x \rfloor / k, & \text{otherwise.} \end{cases}$$

Figure 3.8 shows this function for $k \in \{1, 2\}$. □

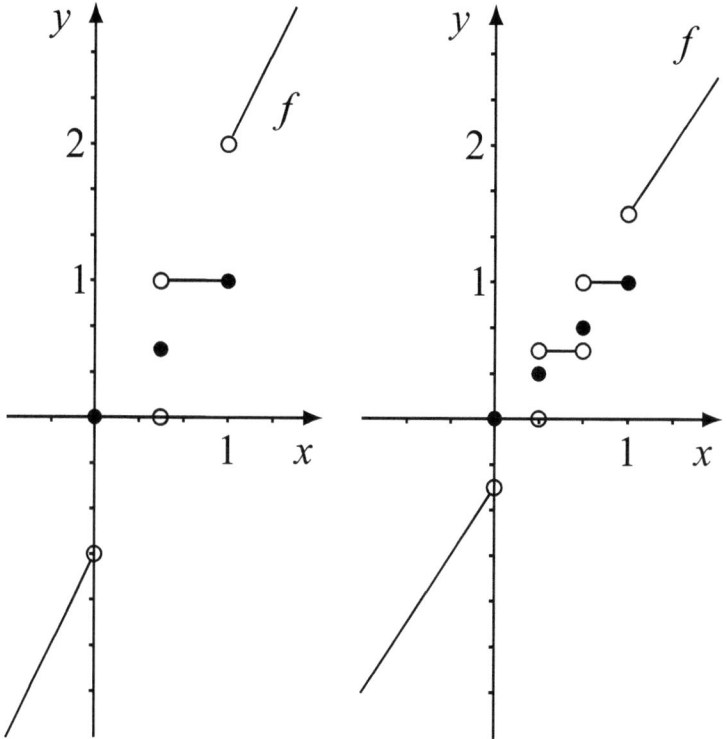

Fig. 3.8 Extending $f_{FS,1}(\cdot; k)$ to a general MDFF f for $k \in \{1, 2\}$ by Proposition 3.12

3.6.4 Examples

3.6.4.1 Based on Method I

Proposition 3.10 provides an approach to construct maximal general dual-feasible functions from classical ones. Since the use of $b > b_0$ never yields extreme maximal dual-feasible functions, one should know the mentioned parameter b_0, whenever it is possible.

If one is given a function $g : [0, 1] \to [0, 1]$ and shall check if g is a maximal dual-feasible function, particularly if g is superadditive, then often the function (3.13) (p. 74) must be tested for nonnegativity, sometimes with long or complex case distinctions. While in this analysis the infimum of h is explored, one can often also find the supremum without much more effort. Hence, the value of b_0 is found almost directly as a side effect of the superadditivity proof.

As an example, we will explore the function $f_{DG,1}$ with the parameters $C \in \mathbb{R} \setminus \mathbb{N}$, $C > 1$ and $k \in \mathbb{N}$ and $k \geq \frac{1}{\text{frac}(C)}$, such that $k \geq 2$. However, the restrictions $x, y \in [0, 1]$ and $x + y \leq 1$ are dropped, because they would not simplify the analysis and $f_{DG,1}$ is a maximal general dual-feasible function, such that all expressions will remain defined. The only negative effect of this relaxation might be an overestimation of b_0. To avoid a confusion with the proof of Proposition 3.9 (p. 68), the identifiers h and d of that proof are kept, and hence h refers here to the auxiliary function (3.8) and not to the function (3.13).

Without loss of generality, we assume that $\text{frac}(Cx) \leq \text{frac}(Cy)$. The first case in the proof of Proposition 3.9 was $\text{frac}(Cy) \leq \text{frac}(C)$, yielding

$$d = \lfloor Cx + Cy \rfloor - \lfloor Cx \rfloor - \lfloor Cy \rfloor + h\left(\frac{\text{frac}(Cx + Cy) - \text{frac}(C)}{1 - \text{frac}(C)}\right).$$

If $\text{frac}(Cx) + \text{frac}(Cy) < 1$, then

$$d = h\left(\frac{\text{frac}(Cx + Cy) - \text{frac}(C)}{1 - \text{frac}(C)}\right) < 1,$$

because the argument of h is below 1. If $\text{frac}(Cx) + \text{frac}(Cy) \geq 1$, then

$$\text{frac}(Cx + Cy) - \text{frac}(C) = \text{frac}(Cx) + \text{frac}(Cy) - 1 - \text{frac}(C) < 0$$

and $d = 1$. This latter situation requires $\text{frac}(C) \geq \frac{1}{2}$. Hence, if $\text{frac}(C) \geq \frac{1}{2}$, then setting $x := y := \frac{\text{frac}(C)}{C}$ will yield $d = 1$. Since $x, y \in [0, 1]$ and

$$x + y = \frac{2\text{frac}(C)}{C} = \frac{2\text{frac}(C)}{\lfloor C \rfloor + \text{frac}(C)} < 1,$$

3.6 Building Maximal General Dual-Feasible Functions

the value $d = 1$ is not an overestimation. It yields $b_0 = 1 + \frac{1}{|C|}$ for the case $\text{frac}(C) \geq \frac{1}{2}$.

Assume in the following that $\text{frac}(C) < \frac{1}{2}$. In that case, we have $k \geq 3$. We want to explore the maximal possible value of d under this condition. The following further cases arise:

1. $\text{frac}(Cx) \leq \text{frac}(C) < \text{frac}(Cy)$ yields two subcases.
   If $\text{frac}(Cx) + \text{frac}(Cy) < 1$, then

$$d = h\left(\frac{\text{frac}(Cx+Cy) - \text{frac}(C)}{1 - \text{frac}(C)}\right) - h\left(\frac{\text{frac}(Cy) - \text{frac}(C)}{1 - \text{frac}(C)}\right)$$

$$\leq h\left(\frac{\text{frac}(Cy)}{1 - \text{frac}(C)}\right) - h\left(\frac{\text{frac}(Cy) - \text{frac}(C)}{1 - \text{frac}(C)}\right)$$

$$\leq \left(\left\lfloor (k-1) \times \frac{\text{frac}(Cy)}{1 - \text{frac}(C)} \right\rfloor + 1 - \left\lceil (k-1) \times \frac{\text{frac}(Cy) - \text{frac}(C)}{1 - \text{frac}(C)} \right\rceil\right) / k$$

$$\leq \left(1 + (k-1) \times \frac{\text{frac}(C)}{1 - \text{frac}(C)}\right) / k$$

$$< 1$$

due to the monotonicity of h, the inequalities (3.9) and $\text{frac}(C) < \frac{1}{2}$.
   If $\text{frac}(Cx) + \text{frac}(Cy) \geq 1$, then

$$d = 1 + h\left(\frac{\text{frac}(Cx) + \text{frac}(Cy) - 1 - \text{frac}(C)}{1 - \text{frac}(C)}\right) - h\left(\frac{\text{frac}(Cy) - \text{frac}(C)}{1 - \text{frac}(C)}\right).$$

Since $\text{frac}(Cx) \leq \text{frac}(C)$, it follows that

$$\text{frac}(Cx) + \text{frac}(Cy) - 1 - \text{frac}(C) \leq \text{frac}(Cy) - 1 < 0,$$

and hence

$$h\left(\frac{\text{frac}(Cx) + \text{frac}(Cy) - 1 - \text{frac}(C)}{1 - \text{frac}(C)}\right) = 0.$$

Therefore, in this case d becomes maximal, if $\text{frac}(Cy)$ is minimized, i.e. for $\text{frac}(Cy) = 1 - \text{frac}(C)$, yielding

$$d = 1 - h\left(1 - \frac{\text{frac}(C)}{1 - \text{frac}(C)}\right)$$

$$= h\left(\frac{\text{frac}(C)}{1 - \text{frac}(C)}\right)$$

$$\leq 1 - \frac{1}{k}.$$

2. The case where $\mathrm{frac}(Cx) > \mathrm{frac}(C)$ and $\mathrm{frac}(Cx) + \mathrm{frac}(Cy) < 1$ yields

$$d = h\left(\frac{\mathrm{frac}(Cx) + \mathrm{frac}(Cy) - \mathrm{frac}(C)}{1 - \mathrm{frac}(C)}\right) - h\left(\frac{\mathrm{frac}(Cx) - \mathrm{frac}(C)}{1 - \mathrm{frac}(C)}\right)$$
$$- h\left(\frac{\mathrm{frac}(Cy) - \mathrm{frac}(C)}{1 - \mathrm{frac}(C)}\right).$$

Hence,

$$dk \leq \left\lfloor (k-1) \times \frac{\mathrm{frac}(Cx) + \mathrm{frac}(Cy) - \mathrm{frac}(C)}{1 - \mathrm{frac}(C)} \right\rfloor + 1$$
$$- \left\lceil (k-1) \times \frac{\mathrm{frac}(Cx) - \mathrm{frac}(C)}{1 - \mathrm{frac}(C)} \right\rceil - \left\lceil (k-1) \times \frac{\mathrm{frac}(Cy) - \mathrm{frac}(C)}{1 - \mathrm{frac}(C)} \right\rceil$$
$$\leq \frac{k-1}{1 - \mathrm{frac}(C)} \times (\mathrm{frac}(Cx) + \mathrm{frac}(Cy) - \mathrm{frac}(C) - \mathrm{frac}(Cx) + \mathrm{frac}(C)$$
$$- \mathrm{frac}(Cy) + \mathrm{frac}(C)) + 1$$
$$= (k-1) \times \frac{\mathrm{frac}(C)}{1 - \mathrm{frac}(C)} + 1,$$

and therefore again $d \leq 1 - \frac{1}{k}$.
3. The case where $\mathrm{frac}(Cx) > \mathrm{frac}(C)$ and $\mathrm{frac}(Cx) + \mathrm{frac}(Cy) \geq 1$ yields

$$d = 1 + h\left(\frac{\mathrm{frac}(Cx) + \mathrm{frac}(Cy) - 1 - \mathrm{frac}(C)}{1 - \mathrm{frac}(C)}\right) - h\left(\frac{\mathrm{frac}(Cx) - \mathrm{frac}(C)}{1 - \mathrm{frac}(C)}\right)$$
$$- h\left(\frac{\mathrm{frac}(Cy) - \mathrm{frac}(C)}{1 - \mathrm{frac}(C)}\right).$$

If $\mathrm{frac}(Cx) + \mathrm{frac}(Cy) > 1 + \mathrm{frac}(C)$, then

$$dk \leq k + \left\lfloor (k-1) \times \frac{\mathrm{frac}(Cx) + \mathrm{frac}(Cy) - 1 - \mathrm{frac}(C)}{1 - \mathrm{frac}(C)} \right\rfloor + 1$$
$$- \left\lceil (k-1) \times \frac{\mathrm{frac}(Cx) - \mathrm{frac}(C)}{1 - \mathrm{frac}(C)} \right\rceil - \left\lceil (k-1) \times \frac{\mathrm{frac}(Cy) - \mathrm{frac}(C)}{1 - \mathrm{frac}(C)} \right\rceil$$
$$\leq k + 1 + \frac{k-1}{1 - \mathrm{frac}(C)} \times (\mathrm{frac}(Cx) + \mathrm{frac}(Cy) - 1 - \mathrm{frac}(C) - \mathrm{frac}(Cx)$$
$$+ \mathrm{frac}(C) - \mathrm{frac}(Cy) + \mathrm{frac}(C))$$
$$= k + 1 - (k-1)$$
$$= 2,$$

3.6 Building Maximal General Dual-Feasible Functions

and hence $d \leq \frac{2}{k}$.

If $\text{frac}(Cx) + \text{frac}(Cy) \leq 1 + \text{frac}(C)$, then

$$h\left(\frac{\text{frac}(Cx) + \text{frac}(Cy) - 1 - \text{frac}(C)}{1 - \text{frac}(C)}\right) = 0,$$

and hence

$$d = 1 - h\left(\frac{\text{frac}(Cx) - \text{frac}(C)}{1 - \text{frac}(C)}\right) - h\left(\frac{\text{frac}(Cy) - \text{frac}(C)}{1 - \text{frac}(C)}\right)$$

$$\leq \frac{k-2}{k}.$$

For the sake of shortness, further details are omitted.

The final result

$$b_0 = 1 + \frac{1}{\lfloor C \rfloor}$$

for $f_{DG,1}(\cdot; C, k)$ with $\text{frac}(C) \geq \frac{1}{2}$, and a smaller b_0 in the case $0 < \text{frac}(C) < \frac{1}{2}$ can be compared with the according b_0 for the function $f_{BJ,1}(\cdot; C)$, which was defined as a classical maximal dual-feasible function in Chap. 2 (see (2.4) at page 25). The latter is also a general maximal dual-feasible function as described in Example 3.1 on page 52. One gets similarly for this function the result

$$b_0 = 1 + \min\left\{1, \frac{\text{frac}(C)}{1 - \text{frac}(C)}\right\} / \lfloor C \rfloor,$$

which is in the case $\text{frac}(C) \geq \frac{1}{2}$ the same value as for $f_{DG,1}$.

3.6.4.2 Based on Method II

Proposition 3.11 requires a Lipschitz-continuous maximal dual-feasible function. This prerequisite is fulfilled by $f_{BJ,1}(\cdot; C)$ for all parameter values $C \geq 1$. This function is one of the strongest known classical dual-feasible functions. Its largest slope, namely

$$t_1 := \frac{C}{\lfloor C \rfloor \times (1 - \text{frac}(C))},$$

is the smallest valid Lipschitz-constant. That can be seen in the term $\frac{1}{\lfloor C \rfloor} \times \frac{\text{frac}(Cx)}{1 - \text{frac}(C)}$ in the definition of $f_{BJ,1}$ in the intervals with slope, i.e. where $\text{frac}(Cx) > \text{frac}(C)$. The additional additive terms in the definition of the function do not influence the

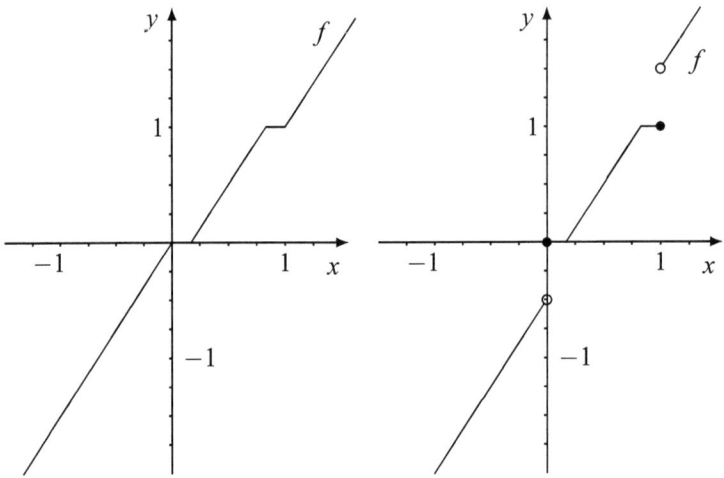

Fig. 3.9 Function f obtained from $f_{BJ,1}$ according to Proposition 3.11 for $p = 1$ and $p = t$

slope. Hence, for any $t \geq t_1$ and $p \in [1, t]$, a general maximal dual-feasible function $f : \mathbb{R} \to \mathbb{R}$ can be obtained by Proposition 3.11, namely

$$f(x) := \begin{cases} tx + 1 - p, & \text{if } x < 0, \\ tx + p - t, & \text{if } x > 1, \\ \left(\lfloor Cx \rfloor + \max\left\{0; \frac{\text{frac}(Cx) - \text{frac}(C)}{1 - \text{frac}(C)}\right\} \right) / \lfloor C \rfloor, & \text{otherwise.} \end{cases}$$

Example 3.14 Figure 3.9 shows the resulting functions, when Proposition 3.11 is used with $f_{BJ,1}(\cdot; \frac{6}{5})$, $t := t_1 = \frac{3}{2}$ and $p \in \{1, t\}$.

In the left case, one has for all $x > 0$ the strict inequality $f(x) < tx$ with

$$t = \sup_{x > 0} \frac{f(x)}{x} = \lim_{x \to \infty} \frac{f(x)}{x},$$

while in the right case it holds that

$$t < -f(-1)$$

(see also Proposition 3.4, p. 62).

□

3.6.4.3 Based on Method III

Proposition 3.12 can be used for every maximal dual-feasible function $g : [0, 1] \to [0, 1]$ due to the weaker prerequisites. It is only necessary to choose

$$t \geq t_0 := \sup\{g(x)/x : 0 < x < 1\}.$$

The value of t_0 is calculated in the sequel for $f_{DG,1}(\cdot; C, k)$ with $C \in \mathbb{R} \setminus \mathbb{N}$, $C > 1$ and $k \in \mathbb{N}$ with $k \geq \frac{1}{\mathsf{frac}(C)}$, such that $k \geq 2$.

Since $f_{DG,1}$ is a staircase function, only the right limits

$$\lim_{y \downarrow x} \frac{f_{DG,1}(y)}{y}$$

at discontinuities may play a role. Discontinuities arise if $Cx \in \mathbb{N} \setminus \{0\}$ or

$$(k-1) \times \frac{\mathsf{frac}(Cx) - \mathsf{frac}(C)}{1 - \mathsf{frac}(C)} \in \mathbb{N}.$$

Let $a := \lfloor Cx \rfloor$ and

$$b := (k-1) \times \frac{\mathsf{frac}(Cx) - \mathsf{frac}(C)}{1 - \mathsf{frac}(C)}.$$

In the first case, one gets

$$f_{DG,1}(x) = a/\lfloor C \rfloor \text{ and } x = a/C,$$

such that

$$t \geq \frac{C}{\lfloor C \rfloor}. \tag{3.15}$$

The second case is $Cx \notin \mathbb{N}$ and $b \in \mathbb{N}$, yielding

$$\lim_{y \downarrow x} f_{DG,1}(y) = \left(a + \frac{b+1}{k}\right)/\lfloor C \rfloor,$$

$b \in \{0, \ldots, k-2\}$, $a < Cx < a + 1$ and

$$b = \frac{k-1}{1 - \mathsf{frac}(C)} \times (Cx - a - \mathsf{frac}(C)).$$

Hence, we have

$$x = \left(\frac{b \times (1 - \text{frac}(C))}{k-1} + a + \text{frac}(C)\right)/C$$

and

$$t \geq \frac{(a + \frac{b+1}{k})/\lfloor C \rfloor}{\left(\frac{b \times (1-\text{frac}(C))}{k-1} + a + \text{frac}(C)\right)/C}. \quad (3.16)$$

Between the two lower bounds (3.15) and (3.16) for t, we keep the largest one. Since $\text{frac}(C) \times k \geq 1$ and $b \leq k - 2$, the following equivalent inequalities are obtained:

$$\text{frac}(C) \times (k^2 - k - bk) \geq k - b - 1$$

$$bk - bk \times \text{frac}(C) + \text{frac}(C) \times (k^2 - k) \geq (k-1) \times (b+1)$$

$$\frac{b \times (1 - \text{frac}(C))}{k-1} + \text{frac}(C) \geq \frac{b+1}{k}$$

$$\frac{b \times (1 - \text{frac}(C))}{k-1} + a + \text{frac}(C) \geq a + \frac{b+1}{k}$$

$$1 \geq \frac{a + \frac{b+1}{k}}{\frac{b \times (1-\text{frac}(C))}{k-1} + a + \text{frac}(C)}$$

Finally, the bound (3.15) is at least as large as (3.16), such that the result is

$$t_0 = \frac{C}{\lfloor C \rfloor}.$$

The same result holds for $f_{BJ,1}$. The application of Proposition 3.12 to $f_{DG,1}$ and $f_{BJ,1}$ for $C = \frac{12}{5}$ (and $k := \lceil \frac{1}{\text{frac}(C)} \rceil = 3$ in the case of $f_{DG,1}$) with $t = t_0 = \frac{6}{5}$ is illustrated in Fig. 3.10.

3.7 Related Literature

The extension of dual-feasible functions to more general domains is recent. The first contributions in this field are due to Rietz et al. (2012b); Rietz et al (2014); Rietz et al. (2015). In Rietz et al. (2012b), the authors discussed the first results of the extension of dual-feasible functions to the domain of real numbers. In Rietz et al (2014), they explored the properties of these general dual-feasible functions. The proof of Lemma 3.2 can be found in this paper, while Lemma 3.3 is an extension of Lemma 3 of Rietz et al. (2010) to domain and range $\mathbb{R}$. The proofs of the other

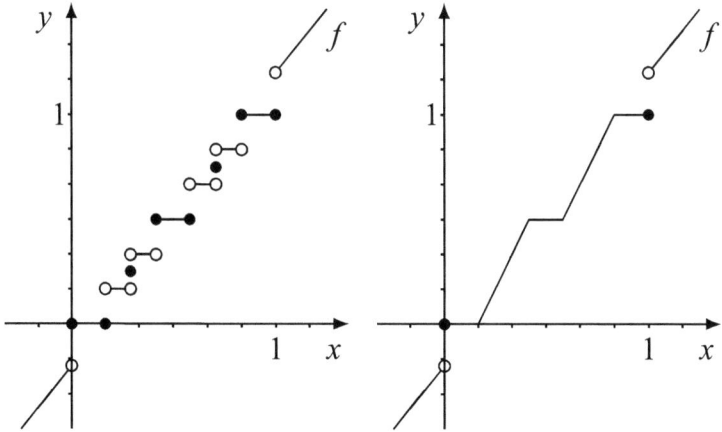

Fig. 3.10 Applying Proposition 3.12 to $f_{DG,1}(\cdot; \frac{12}{5}, 3)$ and $f_{BJ,1}(\cdot; \frac{12}{5})$ for $t = \frac{6}{5}$

properties described in Sect. 3.4 can also be found in Rietz et al (2014). The methods for constructing maximal general dual-feasible functions were first discussed in Rietz et al. (2015).

3.8 Exercises

1. Prove the validity of the necessary conditions of Proposition 3.1.

2. Let $f : \mathbb{R} \to \mathbb{R}$ be a general dual-feasible function with $f(x_0) > 0$ for a certain $x_0 \in \mathbb{R}$. Show that $f(x) < 0$ for all $x < 0$.

3. Find the error in the following reasoning.
    Suppose that someone states (falsely) that every maximal general dual-feasible function $f : \mathbb{R} \to \mathbb{R}$ obeys the symmetry rule (3.3):
    "The symmetry (3.3) holds, if and only if $f(x) + f(1-x) = 1$ for all x. Due to the defining condition on dual-feasible functions, we have $f(x) + f(1-x) \leq 1$, for all x. To show $f(x) + f(1-x) \leq 1$, for all $x \neq 1/2$, we define the function $h : \mathbb{R} \to \mathbb{R}$ as

$$h(x) := \begin{cases} f(x), & \text{if } x \neq x_1, \\ 1 - f(1 - x_1), & \text{otherwise.} \end{cases}$$

Suppose there is an $x_1 > 1/2$ with $f(x_1) + f(1 - x_1) < 1$. Then, h must hurt the defining condition on dual-feasible functions, since f is a maximal dual-feasible function and $h(x) \geq f(x)$, for all x. That requires that there are $n \in \mathbb{N}$ and $x_2, \ldots, x_n \in \mathbb{R}$ with

$$\sum_{i=1}^{n} x_i \leq 1 \text{ and } \sum_{i=1}^{n} h(x_i) > 1,$$

i.e.

$$\sum_{i=2}^{n} f(x_i) > f(1-x_1),$$

but f is superadditive and monotone, implying

$$\sum_{i=2}^{n} f(x_i) \le f(\sum_{i=2}^{n} x_i) \le f(1-x_1).$$

That contradiction proves that the assumed $x_1 > 1/2$ with $f(x_1) + f(1-x_1) < 1$ does not exist. Similar considerations yield $f(1/2) = 1/2$."

4. Consider an instance of the one-dimensional cutting stock problem with a stock length L equal to 122, and six different item lengths $\mathbf{l} = (62, 61, 50, 30, 20, 12)^\top$ with corresponding demands equal to $\mathbf{b} = (2, 1, 1, 1, 2, 4)^\top$.

   (a) Calculate the material bound $z_M := \frac{\mathbf{l}^\top \mathbf{b}}{L}$.
   (b) Provide a feasible integer solution such that the items are cut from no more than $\lceil z_M \rceil + 1$ pieces of the initial material.
   (c) Suppose that we may use once one unit length more in the initial material, i.e. one pattern may use the length 123 instead of 122. Provide an optimal integer solution for this case.
   (d) Calculate lower bounds for the given instance (both without and with the extra length according to task (c)) using the function $f_{BJ,1}$ for all parameter values $C \in \left\{ \frac{L}{\ell_i}, \frac{L}{L-\ell_i} \right\}$. What is the maximum among these bounds for the two cases?

5. Which of the following statements is true?

   (a) Any general dual-feasible function is the sum of a linear and a bounded function.
   (b) Any general dual-feasible functions is bounded from above by a linear function.

6. Let $\mathscr{F}$ be the set of all functions $f : \mathbb{R} \to \mathbb{R}$ with the following two properties:

   - For any finite set $\{x_i \in \mathbb{R} : i \in I\}$ of real numbers, the implication (3.2) holds, i.e., $\sum_{i \in I} x_i \le 0 \implies \sum_{i \in I} f(x_i) \le 0$.
   - Any function $g : \mathbb{R} \to \mathbb{R}$ with $g(x) \ge f(x)$ for all $x \in \mathbb{R}$, which obeys a similar implication as (3.2), is necessarily identical to f, such that f is maximal in this sense.

Let $\mathscr{L}$ be the set of linear functions with nonnegative slope, i.e. $\mathscr{L}$ is the set of functions $f : \mathbb{R} \to \mathbb{R}$ with $f(x) = cx$, where $c \ge 0$ is a constant. Show that $\mathscr{F} = \mathscr{L}$.

3.8 Exercises

7. Let $k \in \mathbb{N} \setminus \{0\}$ be a constant. Define the following functions $f_0, \ldots, f_3 : \mathbb{R} \to \mathbb{R}$ by

$$f_0(x) := (1 + \tanh 1) \times x - \max\{0, \tanh x\};$$

$$f_1(x) := \begin{cases} x, & \text{if } (k+1) \times x \in \mathbb{N}, \\ \lfloor (k+1) \times x \rfloor / k, & \text{otherwise}; \end{cases}$$

$$f_2(x) := \begin{cases} x, & \text{if } (k+1) \times x \in \mathbb{Z}, \\ \lfloor (k+1) \times x \rfloor / k, & \text{otherwise}; \end{cases}$$

$$f_3(x) := \begin{cases} \lceil (k+1) \times x - 1 \rceil / k, & \text{if } x \geq 1, \\ x, & \text{if } (k+1) \times x \in \{0, 1, \ldots, k\}, \\ \lfloor (k+1) \times x \rfloor / k, & \text{otherwise}. \end{cases}$$

Which of these functions are a general dual-feasible function? Which are a maximal general dual-feasible function? Which properties of part (a) of Theorem 3.1 do these functions possess?

8. Show using the following function $g : [0, 1] \to [0, 1]$ that Hölder-continuity instead of Lipschitz-continuity is not enough for the construction method II described in Sect. 3.6.2:

$$g(x) := \begin{cases} (1 - \sqrt{1 - 2x})/2, & \text{if } x \leq 1/2, \\ (1 + \sqrt{2x - 1})/2, & \text{otherwise}. \end{cases}$$

Chapter 4
Applications for Cutting and Packing Problems

4.1 Introduction

Dual-feasible functions have been designed specifically for the cutting-stock problem. As shown in Chap. 1, they arise naturally from the dual of the classical formulation of Gilmore and Gomory for this problem. Since many problems can be modeled using a similar formulation, it makes sense to explore the concept of dual-feasible function within a more general class of applications. A first approach is to consider multi-dimensional dual-feasible functions, which can be used to derive lower bounds for the vector packing problem. Here, we also consider different packing problems with more complicated subproblems such as multi-dimensional orthogonal packing and packing with conflicts. Dual-feasible functions can still be derived in these cases.

4.2 Set-Covering Dual-Feasible Functions

In Chap. 2, dual-feasible functions were introduced as functions $f : [0, 1] \rightarrow [0, 1]$, such that for every finite index set I of real numbers $x_i \in [0, 1]$ with $i \in I$, the following implication holds:

$$\sum_{i \in I} x_i \leq 1 \implies \sum_{i \in I} f(x_i) \leq 1 \qquad (4.1)$$

From a linear programming point of view, it means that the function must produce the values related to a dual solution that is valid for any cutting-stock instance. Clearly, when one particular instance is considered, one may want to relax this strong constraint, and generate a dual solution that is valid for this particular instance.

The data dependency can be introduced by choosing a finite ground set I_0, fixing the numbers $x_i \in [0, 1]$ for all $i \in I_0$, and demanding the implication (4.1) only for any subset $I \subseteq I_0$. In this way, the order demands of bin-packing instances may be considered also in the function f as illustrated in the following example.

Example 4.1 Consider an instance of the 1-dimensional bin-packing problem with four item lengths $\ell_1 = 10$, $\ell_2 = 9$, $\ell_3 = 6$ and $\ell_4 = 4$ and the respective order demands $\mathbf{b} := (1, 1, 2, 1)^\top$. The length 6 is demanded twice, while the other lengths are needed only once. The items have to be packed into the minimal number of bins of length $L = 18$.

One wants to define the function f, such that

$$\sum_{i=1}^{4} b_i \times f(\ell_i/L)$$

is a valid lower bound for the optimal objective function value. If f should be a classical dual-feasible function, then among others

$$f(1/2) \le 1/2 \quad \text{and} \quad f(2/9) \le 1/4$$

would necessarily hold. Since we are discussing data dependent dual-feasible functions f, we set

$$I_0 := \{1, \ldots, 5\},$$

and

$$x_1 := 5/9, \; x_2 := 1/2, \; x_3 := x_4 := 1/3, \; x_5 := 2/9.$$

Additionally, setting now

$$f(x_1) := f(x_2) := 2/3 \quad \text{and} \quad f(x_3) := f(x_4) := f(x_5) := 1/3$$

is feasible for the data dependent function, as it can be checked easily. That yields the lower bound $7/3$. This bound could not be found by a classical dual-feasible function, because the continuous relaxation of the given instance allows to use the patterns

$$(1, 0, 0, 2)^\top,$$
$$(0, 2, 0, 0)^\top,$$
$$(0, 0, 3, 0)^\top,$$
$$(1, 0, 1, 0)^\top,$$

each in the quantity $1/2$, such that the bound becomes 2. Only the last pattern was proper, and even this pattern did not use the given length L completely. □

In the next section, we define formally this concept of *data-dependent set-covering dual-feasible function*, i.e. functions that are designed for a specific instance.

4.2.1 Data-Dependent Dual-Feasible Functions

We consider a ground set $I = \{1, \ldots, n\}$ of elements to cover b_i times each. Let P be a given polyhedral subproblem, and for a given instance D_P of P, let $C(D_P)$ be the finite set of extremal solutions of subproblem P applied to this instance. For any extreme solution (pattern) c of $C(D_P)$, $a_{ic} \in \mathbb{N}$ ($0 \leq a_{ic} \leq b_i$) is the number of times element i is covered in subproblem solution c, and v_c the cost of subproblem solution c. For most bin-packing problems, where the goal is to minimize the number of bins used, the value v_c is equal to 1 for all patterns c. We define λ_c the variables indicating the number of times subproblem solution c is used in the solution.

$$\min \sum_{c \in C(D_P)} v_c \lambda_c \quad (4.2)$$

$$\text{s.t.} \sum_{c \in C(D_P)} a_{ic} \lambda_c \geq b_i, \quad \forall i \in I \quad (4.3)$$

$$\lambda_c \geq 0, \quad \forall c \in C(D_P) \quad (4.4)$$

Now we give a definition of data-dependent set-covering dual-feasible function.

Definition 4.1 Let I be a set of items and D_P an instance of the problem P. Let $C(D_P)$ be the set of extremal solutions of subproblem P related to the instance D_P. A data-dependent set-covering dual-feasible function (SC-DDFF) related to the instance D_P is a mapping g defined from I to $\mathbb{R}_+$ such that

$$c \in C(D_P) \implies \sum_{i \in I} a_{ic} \times g(i) \leq v_c$$

For two different instances of the same problem, a SC-DDFF can be valid for one and not for the other. Unlike classical dual-feasible functions, set-covering dual-feasible functions apply to indices i instead of the sizes ℓ_i. This is due to the fact that in the cutting-stock problem, the size ℓ_i of an item is sufficient to characterize the element, whereas in a more general context, elements may be more complex (several dimensions, a vertex in a graph, for example). In this formalism, geometric constraints of packing applications are modeled as a set of feasible patterns (the set of instance vectors). Practically speaking, being able to characterize this (possibly exponential size) set without enumerating all its elements is crucial.

4.2.2 Data-Independent Dual-Feasible Functions

In certain specific cases, such as the cutting-stock problem, it is possible to define functions that can be applied to any instance. This can be defined properly when the elements i to cover in model (4.2)–(4.4) are characterized in the subproblem by a unique real vector $\mathbf{y}_i \in \mathbb{R}^k$ for a given k. We thus define the notion of set-covering dual-feasible function.

Definition 4.2 Let k be an integer value, P a subproblem, I a set of elements, each element i characterized by a unique real vector $\mathbf{y}_i \in \mathbb{R}^k$, and $C(\mathscr{D}_P)$ the set of **all** possible extremal solutions c related to **any data** of the given subproblem P and v_c defined as above. A set-covering dual-feasible function (SC-DFF) for P is an application f defined from $\mathbb{R}^k$ to $\mathbb{R}_+$ such that

$$c \in C(\mathscr{D}_P) \Longrightarrow \sum_{i \in I} a_{ic} f(\mathbf{y}_i) \leq v_c.$$

Note that any SC-DFF is a data-dependent SC-DFF for any instance.

Obviously, classical dual-feasible functions designed for the cutting-stock problem are special cases of set-covering dual-feasible functions, where $k = 1$, $v_c = 1$, and subproblem P is the knapsack problem. From now on, we will name cutting stock dual-feasible functions (CS-DFF) the classical dual-feasible functions, while the acronym CS-MDFF will be used for maximal CS-DFF.

4.2.3 General Properties

If one is able to compute set-covering dual-feasible functions for a given problem, a fast lower bound is directly obtained.

Proposition 4.1 (Lower Bound Property for SC-DFF) *Let D_P be an instance of problem P that is modeled as a set-covering problem, I the ground set of elements to cover, and $b_i, i \in I$ the number of times each element has to be covered, and $\mathbf{y}_i$ the description of element i in the subproblem. If f is a SC-DFF, then $\sum_{i \in I} b_i \times f(\mathbf{y}_i)$ is a valid lower bound for model (4.2)–(4.4) applied to instance D_P.*

Proposition 4.2 (Lower Bound Property for SC-DDFF) *Let D_P be an instance of problem P that is modeled as a set-covering problem, I the ground set of elements to cover, and $b_i, i \in I$ the number of times each element has to be covered. If f is a*

SC-DDFF dependent on instance D_P, then $\sum_{i \in I} b_i \times f(i)$ is a valid lower bound for model (4.2)–(4.4) applied to instance D_P.

Once a set-covering dual-feasible function is designed, it is possible to generate a large number of other set-covering by applying a cutting-stock dual-feasible function to the values obtained.

Proposition 4.3 (Composition of SC-DFF and CS-DFF) *The composition of a SC-DFF g and a CS-DFF f is a SC-DFF, i.e. $f(g(\cdot))$ is a SC-DFF.*

In the remainder of this chapter we give several examples of dual-feasible functions for different hard combinatorial problems.

4.3 Vector Packing Dual-Feasible Functions

4.3.1 Basic Definition

We now consider the *m*-dimensional vector packing problem (*m*D-VPP). That is a multi-dimensional version of the cutting-stock problem where all dimensions are independent, for instance one dimension is a geometric length and another dimension a time or a weight. In order to simplify the presentation, all sizes are assumed to be normalized, such that the bins become *m*-dimensional unit cubes.

The relation signs $\leq$, $\geq$, $<$ and $>$ will be used for vectors if the relation is componentwise true. For example, $\mathbf{s} \leq \mathbf{t}$ means $s_i \leq t_i$, $i = 1, \ldots, m$. Furthermore, let $\mathbf{w} := (1, 1, \ldots, 1)^\top \in \mathbb{R}^m$, and $\mathbf{o} := (0, 0, \ldots, 0)^\top \in \mathbb{R}^m$.

Problem 4.1 (Vector Packing, *m*D-VPP) An instance $D := (I; \mathbf{L}; \mathbf{b})$ of the *m*D-VPP consists in a set $I = \{1, 2, \ldots, n\}$ of n items, whose sizes are given in the matrix $\mathbf{L}$, and the vector $\mathbf{b} = (b_1, \ldots, b_n)^\top \in (\mathbb{N} \setminus \{0\})^n$ of order demands. In $\mathbf{L} = (l_{11}, l_{12}, \ldots, l_{1m}; \ldots; l_{n1}, l_{n2}, \ldots, l_{nm}) \in [0, 1]^{n \times m}$, the *i*th row-vector is $\mathbf{l}_i$.

The *m*D-VPP consists in partitioning the set of items into a minimum number of subsets such that the items in each subset fit into a bin, i.e. in no dimension the sum of the sizes exceeds 1 for any subset. Hence, a pattern $\mathbf{a} \in \mathbb{N}^n$ is feasible, if the capacity constraints hold on all the *m* dimensions, i.e.

$$\mathbf{L}^\top \mathbf{a} \leq \mathbf{w}.$$

Example 4.2 Consider an instance of the 2-dimensional vector packing problem with $n = 3$ items of sizes

$$\mathbf{L}^\top = \begin{pmatrix} 7/8 & 1/4 & 1/16 \\ 3/8 & 3/8 & 5/16 \end{pmatrix},$$

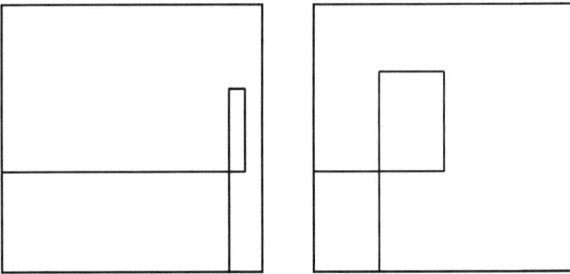

Fig. 4.1 Optimal solution of a 2-dimensional vector packing problem instance (Example 4.2)

and demands $\mathbf{b} = (1, 2, 1)^\top$, such that the second item is demanded twice. The only optimal solution consists in using the two feasible patterns

$$\mathbf{a}^1 = (1, 0, 1)^\top \quad \text{and} \quad \mathbf{a}^2 = (0, 2, 0)^\top,$$

each in the quantity 1. This solution is illustrated in Fig. 4.1. The remaining space in the two bins is

$$(1/16, 5/16)^\top \quad \text{and} \quad (1/2, 1/4)^\top,$$

such that one additional instance of the third item would have fit into the first bin.

This example demonstrates also the following difference between the vector packing problem and the 1-dimensional cutting-stock problem. In the latter, the material bound is generally weak, but it is always above half of the optimal objective function value, and it can be used for some theoretical estimations. However, in the vector packing problem, the material bound becomes even weaker, because it is only subadditive and no longer additive, when more items are added. The material bound would yield in this example $\frac{15}{16}$ usage of the first bin in the first dimension, and $\frac{3}{4}$ for the second bin in the second dimension, while the total material bound is only $\frac{23}{16} < \frac{15}{16} + \frac{3}{4}$. □

A straightforward way of computing lower bounds for the mD-VPP ($m > 1$) would be to consider each dimension of the multi-dimensional problem independently and to compute a lower bound for each of the related m instances of the 1-dimensional bin-packing problem separately, for example by applying a dual-feasible function to the obtained data. However, this may lead to arbitrarily bad results. Consider as an example an instance, where each item i has a size equal to 1 on dimension i and ε on the other dimensions, $\varepsilon > 0$ sufficiently small. Any bound based on that decomposition into m independent 1-dimensional problems yields the optimal value 2 for all the m problems, such that finally at most two is obtained as lower bound, although one needs m bins to pack all the items. By increasing the value of m, the ratio between the optimal objective function value and the obtained lower bound can become arbitrarily bad.

4.3 Vector Packing Dual-Feasible Functions

To avoid this bad behaviour, the m dimensions must be considered simultaneously. That applies also to the functions to be defined. Therefore, we apply the concept of set-covering dual-feasible function, where the subproblem is the following multi-dimensional knapsack problem.

Problem 4.2 (Multi-Dimensional Knapsack, mD-KP) An instance $D := (I; \mathbf{L}; \mathbf{b}; \pi)$ of the mD-KP consists in a set I of n items whose sizes are given in the matrix $\mathbf{L}$, and whose profits are given in the vector π, and the vector $\mathbf{b}$ of upper bounds.

The mD-KP can be stated as follows. mD-KP$(I; \mathbf{L}; \mathbf{b}; \pi) = \{\max \sum_{i=1}^{n} \pi_i x_i : \sum_{i=1}^{n} \ell_{ij} x_i \leq 1, j = 1, \ldots, m, 0 \leq x_i \leq b_i, i = 1, \ldots, n\}$.

We now introduce the concept of vector packing dual-feasible functions, where the 1-dimensional domain is now replaced by an m-dimensional one. These functions can be applied directly to mD-VPP instances without separating the m dimensions. Therefore, they may lead to much stronger lower bounds for the mD-VPP. Our aim consists in getting results that are similar to those obtained for the 1-dimensional case.

First, we introduce formally the definitions of (maximal) vector packing dual-feasible functions.

Definition 4.3 A function $f : [0,1]^m \to [0,1]$ is a *vector packing dual-feasible function* (VP-DFF), if for all instances of the mD-VPP and all feasible patterns $\mathbf{a} \in \mathbb{N}^n$ satisfying $\sum_{i=1}^{n} a_i \ell_{id} \leq w_d, d = 1, \ldots, m$, the following inequality holds:

$$\sum_{i=1}^{n} a_i \times f(\mathbf{l}_i^\top) \leq 1.$$

A VP-DFF is called *maximal* if it is not dominated by another VP-DFF, as stated in the following definition.

Definition 4.4 A VP-DFF f is *maximal* (VP-MDFF), if there is no other VP-DFF g with $g(\mathbf{x}) \geq f(\mathbf{x})$ for all $\mathbf{x} \in [0,1]^m$.

Examples of some simple VP-DFF are the projections to the jth coordinate of the argument-vector ($j = 1, \ldots, m$), i.e. $f_j(\mathbf{x}) = x_j$. These functions lead to lower bounds for the mD-VPP which are already known from the 1-dimensional bin-packing problem for the separated m dimensions.

Example 4.3 Consider the instance of Example 4.2 with fourfold demand vector, i.e., now let $\mathbf{b} := (4, 8, 4)^\top$. The simple material bound would yield $23/4 < 6$. Separating the two dimensions would lead to the bound 6 in the first dimension and 7 in the second direction, because one could use twice the pattern $(1, 0, 2)^\top$, such that ten items with size $3/8$ would remain, which would fill five more bins. Finally, the maximum of the bounds is to be taken, yielding the value 7. However, the first item is incompatible with the second one due to the first dimension, such that the remaining demand $(2, 8, 0)^\top$ requires $2 + \lceil 8/2 \rceil = 6$ more bins, i.e. 8.

A bound 8 could be achieved, if we would have e.g. a VP-DFF $f : [0, 1]^2 \to [0, 1]$ with

$$f\left(\frac{7}{8}, \frac{3}{8}\right) = 1, \, f\left(\frac{1}{4}, \frac{3}{8}\right) = \frac{1}{2}, \, f\left(\frac{1}{16}, \frac{5}{16}\right) = 0,$$

because

$$b_1 \times 1 + b_2 \times \frac{1}{2} + b_3 \times 0 = 8.$$

Note that such a function f can be constructed according to Proposition 4.9 (p. 106), which will be introduced later as a general construction principle, with $\mathbf{u} := (\frac{1}{4}, 0)^\top$ and with g being the function (4.8) (p. 100) with parameter $C = 8/3$. □

4.3.2 General Properties of VP-MDFF

We now show that some properties for the 1-dimensional case can be generalized to the multidimensional case, and we give a complete characterization of *maximal* functions for the m-dimensional case. Finally, we show how to build such *maximal* functions from non-*maximal* superadditive ones by forcing symmetry.

The necessary conditions from the 1-dimensional case for a function to be maximal are still valid for the higher-dimensional case. However, it has to be checked how the higher-dimensional case can be described and if stronger sufficient conditions are needed. These ideas led to the following theorems.

Theorem 4.1 *Any VP-MDFF* $f : [0, 1]^m \to [0, 1]$ *has necessarily the following properties:*

1. f is superadditive, i.e. for all $\mathbf{x}, \mathbf{y} \in [0, 1]^m$ with $\mathbf{x} + \mathbf{y} \leq \mathbf{w}$, it holds that

$$f(\mathbf{x} + \mathbf{y}) \geq f(\mathbf{x}) + f(\mathbf{y}); \tag{4.5}$$

2. f is non-decreasing:

$$f(\mathbf{x}) \leq f(\mathbf{y}), \, if \, \mathbf{o} \leq \mathbf{x} \leq \mathbf{y} \leq \mathbf{w};$$

3. f is symmetric, i.e. for all $\mathbf{x} \in [0, 1]^m$, it holds that

$$f(\mathbf{x}) + f(\mathbf{w} - \mathbf{x}) = 1, \tag{4.6}$$

and especially $f(\mathbf{w}) = 1$ and $f(\frac{1}{2}\mathbf{w}) = 1/2$.

Properties (4.5)–(4.6) of Theorem 4.1 are also sufficient conditions for a function $f : [0, 1]^m \to [0, 1]$ to be a VP-MDFF. However, more restricted sufficient conditions

4.3 Vector Packing Dual-Feasible Functions

can be derived too, as stated in Theorem 4.2. This theorem may help to simplify the proofs of maximality in the next sections. Before introducing these new sufficient conditions, first in Lemma 4.1 an additional assertion is described that is useful to prove the maximality of a VP-DFF.

Lemma 4.1 *If a VP-DFF $f : [0, 1]^m \to [0, 1]$ satisfies the symmetry condition (4.6), then f is a VP-MDFF.*

This is obvious since for such a symmetric VP-DFF f, if there is a VP-DFF g such that $g(\mathbf{x}) > f(\mathbf{x})$ for a given $\mathbf{x}$, then $g(\mathbf{w} - \mathbf{x}) < f(\mathbf{w} - \mathbf{x})$ must hold (otherwise $g(\mathbf{x}) + g(\mathbf{w} - \mathbf{x}) > 1$).

Example 4.4 Consider the 2-dimensional case, and let $f : [0, 1]^2 \to [0, 1]$ be the following function:

$$f(\mathbf{x}) := \begin{cases} 0, & \text{if } \mathbf{x} \leq \frac{1}{2}\mathbf{w} \wedge \mathbf{x} \neq \frac{1}{2}\mathbf{w}, \\ 1, & \text{if } \mathbf{x} \geq \frac{1}{2}\mathbf{w} \wedge \mathbf{x} \neq \frac{1}{2}\mathbf{w}, \\ \frac{1}{2}, & \text{otherwise}. \end{cases} \quad (4.7)$$

This function is a VP-DFF, because one has, for any finite set of vectors $\mathbf{x}^1, \ldots, \mathbf{x}^n \in [0, 1]^2$ with $\sum_{i=1}^n \mathbf{x}^i \leq \mathbf{w}$, that $\sum_{i=1}^n f(\mathbf{x}^i) \leq 1$, as it can be checked easily. Additionally, since the symmetry condition holds, f is maximal, and hence it is a VP-MDFF. □

The following theorem restricts the needed sufficient conditions for maximality proofs. The idea came from the 1-dimensional case, and here twice a certain dimension can be chosen, where the weaker sufficient conditions of the 1-dimensional cutting stock problem are applied.

Theorem 4.2 *Given two constants $r, s \in \{1, \ldots, m\}$ and a function $f : [0, 1]^m \to [0, 1]$, the following conditions are sufficient for f to be a VP-MDFF:*

1. *Equation (4.6) is true for all $\mathbf{x} \in [0, 1]^m$ with $x_r \leq 1/2$;*
2. *Inequality (4.5) holds for all $\mathbf{x}, \mathbf{y} \in [0, 1]^m$ with $\mathbf{x} + \mathbf{y} \leq \mathbf{w}$, and $x_s \leq y_s \leq 1/2$ and $x_s + y_s \leq 2/3$.*

The following propositions state that the functions resulting from the convex combination of VP-MDFF or from the composition of a VP-MDFF with a maximal dual-feasible function remain maximal.

Proposition 4.4 *Any convex combination of VP-MDFF is a VP-MDFF.*

Example 4.5 Consider again the 2-dimensional case. The following function $g : [0, 1]^2 \to [0, 1]$ is a VP-MDFF:

$$g(\mathbf{x}) := \frac{x_1 + x_2}{2}.$$

The convex combination of g with the function (4.7) with the weights $\lambda \in [0, 1]$ and $1 - \lambda$, respectively, yields another VP-MDFF $h : [0, 1]^2 \to [0, 1]$, namely

$$h(\mathbf{x}) = \begin{cases} \frac{\lambda}{2} \times (x_1 + x_2), & \text{if } \mathbf{x} \leq \frac{1}{2}\mathbf{w} \wedge \mathbf{x} \neq \frac{1}{2}\mathbf{w}, \\ \frac{\lambda}{2} \times (x_1 + x_2) + 1 - \lambda, & \text{if } \mathbf{x} \geq \frac{1}{2}\mathbf{w} \wedge \mathbf{x} \neq \frac{1}{2}\mathbf{w}, \\ \frac{\lambda}{2} \times (x_1 + x_2) + \frac{1-\lambda}{2}, & \text{otherwise.} \end{cases}$$

□

Proposition 4.5 *The composition of a VP-MDFF f with a MDFF g, i.e. $g(f(\cdot))$, is a VP-MDFF.*

Example 4.6 Let $f : [0, 1]^m \to [0, 1]$ be simply

$$f(\mathbf{x}) := \frac{\|\mathbf{x}\|_1}{m} = \frac{x_1 + \cdots + x_m}{m}$$

and let $g : [0, 1] \to [0, 1]$ be the parameter dependent function

$$g(x) := \begin{cases} \lfloor Cx \rfloor / \lfloor C \rfloor, & \text{if } x < 1/2, \\ 1/2, & \text{if } x = 1/2, \\ 1 - g(1 - x), & \text{otherwise,} \end{cases} \quad (4.8)$$

with $C \in \mathbb{R}$, $C \geq 1$. That yields the composed function

$$g(f(\mathbf{x})) = \begin{cases} \lfloor \frac{C}{m} \times \|\mathbf{x}\|_1 \rfloor / \lfloor C \rfloor, & \text{if } \|\mathbf{x}\|_1 < \frac{m}{2}, \\ 1/2, & \text{if } \|\mathbf{x}\|_1 = \frac{m}{2}, \\ 1 - g(1 - f(\mathbf{x})), & \text{otherwise.} \end{cases}$$

□

We now describe how a VP-MDFF can be built from a superadditive m-dimensional vector function by forcing symmetry. This result generalizes Theorem 2.4.

Proposition 4.6 *Let $f : [0, 1]^m \to [0, 1]$ be a superadditive function, and M be any subset of $[0, 1]^m \setminus \{\frac{1}{2}\mathbf{w}\}$ such that:*

1. *for all $\mathbf{x} \in [0, 1]^m \setminus \{\frac{1}{2}\mathbf{w}\}$, the following equivalence holds:*

$$\mathbf{x} \in M \iff \mathbf{w} - \mathbf{x} \notin M;$$

2. *for any $\mathbf{x}, \mathbf{y} \in M$, it holds that*

$$\mathbf{x} + \mathbf{y} \not\leq \mathbf{w}. \quad (4.9)$$

4.3 Vector Packing Dual-Feasible Functions

The following function $g : [0, 1]^m \to [0, 1]$ built from f is a VP-MDFF:

$$g(\mathbf{x}) := \begin{cases} 1/2, & \text{if } 2\mathbf{x} = \mathbf{w}, \\ 1 - f(\mathbf{w} - \mathbf{x}), & \text{if } \mathbf{x} \in M, \\ f(\mathbf{x}), & \text{otherwise.} \end{cases}$$

Example 4.7 There are various ways to choose the set M in Proposition 4.6, for instance as the union of m parts according to

$$M := [0, 1] \times [0, 1] \times \cdots \times [0, 1] \times \left(\frac{1}{2}, 1\right] \cup$$
$$\cup [0, 1] \times \cdots \times [0, 1] \times \left(\frac{1}{2}, 1\right] \times \left\{\frac{1}{2}\right\} \cup \cdots \cup$$
$$\cup \left(\frac{1}{2}, 1\right] \times \left\{\frac{1}{2}\right\} \times \cdots \times \left\{\frac{1}{2}\right\}.$$

For $m = 2$, this set becomes $[0, 1] \times \left(\frac{1}{2}, 1\right] \cup \left(\frac{1}{2}, 1\right] \times \left\{\frac{1}{2}\right\}$, i.e. the upper half of the unit square, where the border belongs only partially to M. □

Example 4.8 Additionally, M could be chosen for example as follows, where the inner of M is a triangle:

$$M := \{\mathbf{x} \in [0, 1]^2 : x_1 + x_2 > 1\} \cup \{\mathbf{x} \in (\frac{1}{2}, 1] \times [0, 1] : x_1 + x_2 = 1\}.$$

The resulting function obtained by applying Proposition 4.6 to a random superadditive VP-DFF is depicted in Fig. 4.2. □

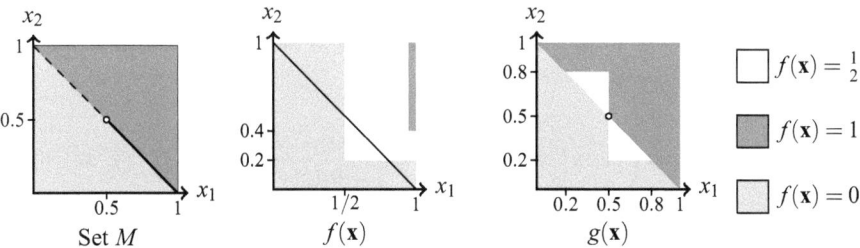

Fig. 4.2 Using Proposition 4.6 to build a VP-DFF. For set M, the *dashed line* and the *point* $(1/2, 1/2)$ do not belong to M, while the *solid line* does

4.3.3 General Classes of VP-MDFF

In this section, several general classes of VP-MDFF are described. When general schemes for generating VP-MDFF are proposed, some specific functions that can be obtained from these schemes are described and analyzed.

4.3.3.1 Class I: Functions Based on Projections into 1-Dimensional Domains

The first set of VP-MDFF is built from the projection of the m-dimensional data into 1-dimensional domains. A formal definition of these VP-MDFF is given in Proposition 4.7.

Proposition 4.7 *Let $g : [0, 1] \to [0, 1]$ be a MDFF and $\mathbf{u} \in \mathbb{R}_+^m$ with $\mathbf{u}^\top \mathbf{w} = 1$. The function $f_I(\cdot; g, \mathbf{u}) : [0, 1]^m \to [0, 1]$ with*

$$f_I(\mathbf{x}; g, \mathbf{u}) := g(\mathbf{u}^\top \mathbf{x})$$

is a VP-MDFF.

Using the MDFF $f_{FS,1}$ in Proposition 4.7 yields the function described in Corollary 4.1.

Corollary 4.1 *Let $\mathbf{v} \in \mathbb{R}_+^m$ such that $\mathbf{v}^\top \mathbf{w} \in \mathbb{N} \setminus \{0, 1\}$. The following function $f_{I,FS,1}(\cdot; \mathbf{v}) : [0, 1]^m \to \mathbb{R}_+$ is a VP-MDFF:*

$$f_{I,FS,1}(\mathbf{x}, \mathbf{v}) := \begin{cases} \frac{\mathbf{v}^\top \mathbf{x}}{\mathbf{v}^\top \mathbf{w}}, & \text{if } \mathbf{v}^\top \mathbf{x} \in \mathbb{N}, \\ \frac{\lfloor \mathbf{v}^\top \mathbf{x} \rfloor}{\mathbf{v}^\top \mathbf{w} - 1}, & \text{otherwise.} \end{cases}$$

Applying Proposition 4.7 with the MDFF $f_{BJ,1}$ leads to the following function.

Corollary 4.2 *Let $\mathbf{v} \in \mathbb{R}_+^m$ be any vector with $\mathbf{v}^\top \mathbf{w} \geq 1$. Then, the function $f_{I,BJ,1}(\cdot; \mathbf{v}) : [0, 1]^m \to \mathbb{R}_+$ with*

$$f_{I,BJ,1}(\mathbf{x}; \mathbf{v}) := \left(\lfloor \mathbf{v}^\top \mathbf{x} \rfloor + \max \left\{ 0, \frac{\text{frac}(\mathbf{v}^\top \mathbf{x}) - \text{frac}(\mathbf{v}^\top \mathbf{w})}{1 - \text{frac}(\mathbf{v}^\top \mathbf{w})} \right\} \right) / \lfloor \mathbf{v}^\top \mathbf{w} \rfloor$$

is a VP-MDFF.

Note that the function $f_{I,BJ,1}(\cdot; \mathbf{v})$ of Corollary 4.2 is only a convex combination of the projections $f_1, \ldots, f_m$ if $\mathbf{v}^\top \mathbf{w} \in \mathbb{N}$.

Example 4.9 To demonstrate how a function f of Class I can be built, consider the simple function $f_{MT,0}(\cdot; \frac{1}{2})$, defined in Formula (2.16), p. 48, applied to each dimension, with different vectors $\mathbf{u}$. We restrict the example to two dimensions.

4.3 Vector Packing Dual-Feasible Functions

If $\mathbf{u} = (1,0)^\top$, then

$$f(\mathbf{x}) = 0, \text{ if } x_1 < \frac{1}{2},$$
$$f(\mathbf{x}) = \frac{1}{2}, \text{ if } x_1 = \frac{1}{2},$$
$$f(\mathbf{x}) = 1, \text{ if } x_1 > \frac{1}{2}.$$

If $\mathbf{u} = \left(\frac{3}{4}, \frac{1}{4}\right)^\top$, then

$$f(\mathbf{x}) = 0, \text{ if } \frac{3}{4}x_1 + \frac{1}{4}x_2 < \frac{1}{2},$$
$$f(\mathbf{x}) = \frac{1}{2}, \text{ if } \frac{3}{4}x_1 + \frac{1}{4}x_2 = \frac{1}{2},$$
$$f(\mathbf{x}) = 1, \text{ if } \frac{3}{4}x_1 + \frac{1}{4}x_2 > \frac{1}{2}.$$

If $\mathbf{u} = \left(\frac{1}{2}, \frac{1}{2}\right)^\top$, then

$$f(\mathbf{x}) = 0, \text{ if } x_1 + x_2 < 1,$$
$$f(\mathbf{x}) = \frac{1}{2}, \text{ if } x_1 + x_2 = 1,$$
$$f(\mathbf{x}) = 1, \text{ if } x_1 + x_2 > 1.$$

The behaviour of these functions, and other examples of parameters, are illustrated in Fig. 4.3. □

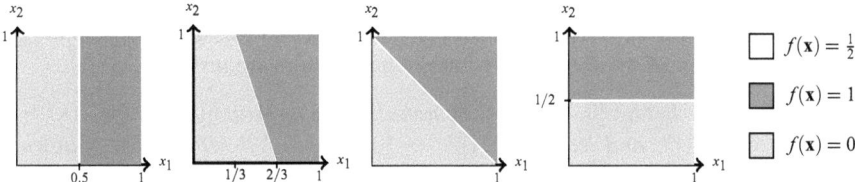

Fig. 4.3 Behaviour of class I functions for respectively $\mathbf{u} = (1,0)^\top, \left(\frac{3}{4}, \frac{1}{4}\right)^\top, \left(\frac{1}{2}, \frac{1}{2}\right)^\top$ and $(0,1)^\top$ when function $f_{MT,0}(L; \frac{1}{2})$ is used

Example 4.10 Recall Example 1.7, p. 14. After scaling the values of the weights and volumes, the 2-dimensional items have the following sizes:

Item	1	2	3	4	Vehicle
Volume (x_1)	0.5	0.75	0.25	0.5	1
Weight (x_2)	0.6	0.4	0.8	0.2	1

Using the function $f_{MT,0}(\cdot;\frac{1}{2})$ for different values of $\mathbf{u}$, the sizes of the items are mapped into the values indicated in the following table:

(x_1,x_2)	$\mathbf{u}=(1,0)^\top$	$\mathbf{u}=(\frac{3}{4},\frac{1}{4})^\top$	$\mathbf{u}=(\frac{1}{2},\frac{1}{2})^\top$	$\mathbf{u}=(\frac{1}{4},\frac{3}{4})^\top$	$\mathbf{u}=(0,1)^\top$
(0.5, 0.6)	$\frac{1}{2}$	1	1	1	1
(0.75, 0.4)	1	1	1	0	0
(0.25, 0.8)	0	0	1	1	1
(0.5, 0.2)	$\frac{1}{2}$	0	0	0	0

Note that the several DFF in the family provide the dual feasible solutions $\hat{\mathbf{u}}_1 = (\frac{1}{2},1,0,\frac{1}{2})^\top$, $\hat{\mathbf{u}}_2 = (1,1,0,0)^\top$, $\hat{\mathbf{u}}_3 = (1,1,1,0)^\top$ and $\hat{\mathbf{u}}_4 = (1,0,1,0)^\top$, indicated in Example 1.7.

As all the demands are equal to 1, the value of the lower bound is equal to the sum of the elements of $\hat{\mathbf{u}}$. The function with $\mathbf{u} = (\frac{1}{2},\frac{1}{2})^\top$ that provides the dual feasible solution $\hat{\mathbf{u}}_3 = (1,1,1,0)^\top$ is the one that yields the best lower bound, equal to 3. □

4.3.3.2 Class II

Some of the ideas of the 1-dimensional MDFF can be adapted for the m-dimensional vector packing problem, for instance, the function which maps small items to zero and large ones to 1, while the other items remain unchanged, can be generalized in the following way. Note that the difficulty in this generalization lies in finding a suitable definition of *small* and *large* items when vectors are involved.

Proposition 4.8 *Let $h : [0,1]^m \to \mathbb{R}$ be non-decreasing with $h(\mathbf{x}) + h(\mathbf{w}-\mathbf{x}) > 0$ for all $\mathbf{x} \in [0,1]^m$, and let $g : [0,1]^m \to [0,1]$ be a VP-MDFF. The following functions $f_{II1}, f_{II2} : [0,1]^m \to [0,1]$ are VP-MDFF:*

$$f_{II1}(\mathbf{x}) := \begin{cases} 0, & \text{if } h(\mathbf{x}) \leq 0 \\ 1, & \text{if } h(\mathbf{w}-\mathbf{x}) \leq 0 \\ g(\mathbf{x}), & \text{otherwise} \end{cases} \quad f_{II2}(\mathbf{x}) := \begin{cases} 0, & \text{if } h(\mathbf{x}) < 0 \\ 1, & \text{if } h(\mathbf{w}-\mathbf{x}) < 0 \\ g(\mathbf{x}), & \text{otherwise} \end{cases}.$$

4.3 Vector Packing Dual-Feasible Functions

Corollary 4.3 *Let* $\mathbf{u} \in \left[0, \frac{1}{2}\right]^m$, *and let* $g : [0, 1]^m \to [0, 1]$ *be a VP-MDFF. The following functions* $f_{II3}(\cdot; g, \mathbf{u}), f_{II4}(\cdot; g, \mathbf{u}) : [0, 1]^m \to [0, 1]$ *are also VP-MDFF:*

$$f_{II3}(\mathbf{x}; g, \mathbf{u}) := \begin{cases} 0, & \text{if } \mathbf{x} \leq \mathbf{u} \text{ and } \mathbf{x} \neq \frac{1}{2}\mathbf{w}, \\ 1, & \text{if } \mathbf{x} \geq \mathbf{w} - \mathbf{u} \text{ and } \mathbf{x} \neq \frac{1}{2}\mathbf{w}, \\ g(\mathbf{x}), & \text{otherwise;} \end{cases}$$

$$f_{II4}(\mathbf{x}; g, \mathbf{u}) := \begin{cases} 0, & \text{if } \mathbf{x} < \mathbf{u}, \\ 1, & \text{if } \mathbf{x} > \mathbf{w} - \mathbf{u}, \\ g(\mathbf{x}), & \text{otherwise.} \end{cases}$$

Corollary 4.4 *Let* $\|\cdot\|_p$ *be an* $\mathscr{L}^p$-*norm in* $\mathbb{R}^m$ *with* $1 \leq p \leq \infty$, *i.e.*

$$\|\mathbf{x}\|_\infty = \max_{r=1,\ldots,m} |x_r| \text{ and } \|\mathbf{x}\|_p = \sqrt[p]{\sum_{r=1}^m |x_r|^p} \text{ for } p < \infty. \quad (4.10)$$

Let $g : [0, 1] \to [0, 1]$ *be a VP-MDFF and* $\varepsilon \in (0, \|\mathbf{w}\|_p/2)$. *The following function* $f_{II5} : [0, 1]^m \to [0, 1]$ *is a VP-MDFF:*

$$f_{II5}(\mathbf{x}) := \begin{cases} 0, & \text{if } \|\mathbf{x}\|_p \leq \varepsilon, \\ 1, & \text{if } \|\mathbf{w} - \mathbf{x}\|_p \leq \varepsilon, \\ g(\mathbf{x}), & \text{otherwise.} \end{cases}$$

Corollary 4.5 *Let* $g : [0, 1]^m \to \mathbb{R}$ *be any non-decreasing function, and let* $r \in \{1, \ldots, m\}$. *The following function* $f_{II6} : [0, 1]^m \to [0, 1]$ *is a VP-MDFF:*

$$f_{II6}(\mathbf{x}) := \begin{cases} 0, & \text{if } 2\mathbf{w}^\top \mathbf{x} < m \text{ and } g(\mathbf{x}) < 0, \\ 1, & \text{if } 2\mathbf{w}^\top \mathbf{x} > m \text{ and } g(\mathbf{w} - \mathbf{x}) < 0, \\ x_r, & \text{otherwise.} \end{cases}$$

Example 4.11 A 2-dimensional example of function $f_{II3}(\cdot; g, \mathbf{u})$ obtained by using Corollary 4.3 is given next. The VP-MDFF $g : [0, 1]^2 \to [0, 1]$ in the left-hand side of Fig. 4.4 is defined by

$$g(\mathbf{x}) := \begin{cases} 0, & \text{if } x_1 < 1/4, \\ 1, & \text{if } x_1 > 3/4, \\ x_1, & \text{otherwise.} \end{cases}$$

Using $\mathbf{u} := (1/3, 1/3)^\top$ results in the function $f_{II3}(\cdot; g, \mathbf{u})$ depicted in the right part of Fig. 4.4. □

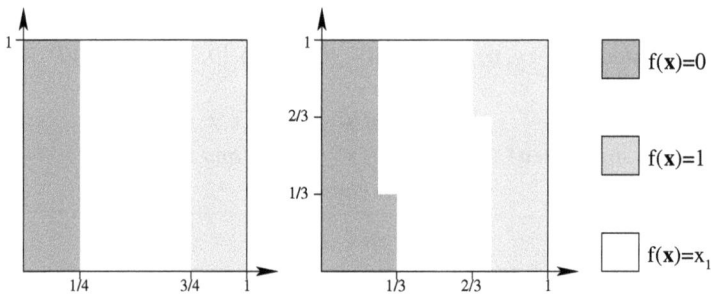

Fig. 4.4 Applying Corollary 4.3 to the *left-hand* VP-MDFF, using $\mathbf{u} = (\frac{1}{3}, \frac{1}{3})^\top$

4.3.3.3 Class III

The following proposition describes another VP-MDFF. First the general function for the m-dimensional case is defined, then a special case for $m = 2$ is given in Corollary 4.6. The rationale behind this function consists in assigning the value 0 (or 1) to items that are very small (respectively large) on some dimensions, unless they are very large (or small) on another dimension. The remaining items are mapped via any other VP-MDFF of appropriate dimension.

Proposition 4.9 *Let* $g : [0,1]^m \to [0,1]$ *be a VP-MDFF and* $\mathbf{u} \in [0, 1/2]^m$. *The following function* $f_{III1} : [0,1]^m \to [0,1]$

$$f_{III1}(\mathbf{x}) := \begin{cases} 0, & \text{if } \exists i \in \{1,\ldots,m\} : x_i < u_i \wedge \not\exists j \in \{1,\ldots,i\} \text{ with } x_j > 1 - u_j, \\ 1, & \text{if } \exists i \in \{1,\ldots,m\} : x_i > 1 - u_i \wedge \not\exists j \in \{1,\ldots,i\} \text{ with } x_j < u_j, \\ g(\mathbf{x}), & \text{otherwise}, \end{cases}$$

is a VP-MDFF.

Corollary 4.6 *The function* $f_{III2}(\cdot; u_1, u_2, r, q) : [0,1]^2 \to [0,1]$ *with* $0 \leq u_1, u_2 \leq 1/2$ *and* $r, q \in \{1, 2\}$, *which is defined as*

$$f_{III2}(\mathbf{x}; u_1, u_2, r, q) := \begin{cases} 1, & \text{if } x_q > 1 - u_1 \text{ or } (x_q \geq u_1 \text{ and } x_{3-q} > 1 - u_2), \\ x_r, & \text{if } 1 - u_1 \geq x_q \geq u_1 \text{ and } 1 - u_2 \geq x_{3-q} \geq u_2, \\ 0, & \text{otherwise}, \end{cases}$$

is a VP-MDFF.

Example 4.12 Consider an example in dimension 2 and choose $r = 1$, $q = 2$, $u_1 = 1/4$ and $u_2 = 1/3$.

$$f_{III2}\left(\mathbf{x}; \frac{1}{4}, \frac{1}{3}, 1, 2\right) := \begin{cases} 1, & \text{if } x_2 > 3/4 \text{ or } (x_2 \geq 1/4 \text{ and } x_1 > 2/3), \\ x_1, & \text{if } 3/4 \geq x_2 \geq 1/4 \text{ and } 2/3 \geq x_1 \geq 1/3, \\ 0, & \text{otherwise}, \end{cases}$$

4.3 Vector Packing Dual-Feasible Functions

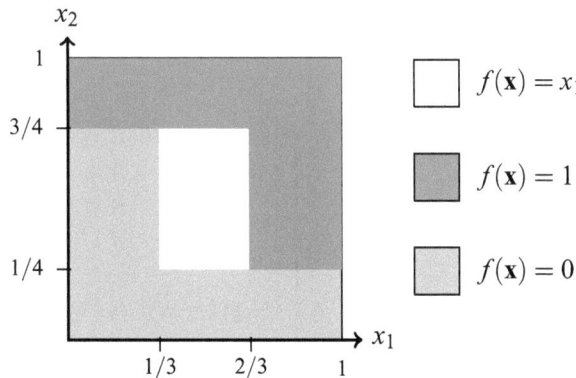

Fig. 4.5 Applying Corollary 4.6 with $r = 1$, $q = 2$, $u_1 = 1/4$ and $u_2 = 1/3$

The obtained function is depicted in Fig. 4.5. □

4.3.3.4 Class IV

In this subsection, another class of VP-MDFF in its general form and a specialization for the 2-dimensional case are described. This VP-MDFF depends on a non-decreasing function g whose properties are stated in the next proposition. How g should be chosen is explained at the end of the section.

Proposition 4.10 *Let $m \in \mathbb{N} \setminus \{0\}$, $k \in (1/3, 1/2]$, $g : [0, 1]^m \to [0, 1]$ be a non-decreasing function with*

$$\mathbf{x}, \mathbf{y} \in [0, 1]^m, \mathbf{x} + \mathbf{y} \geq \mathbf{w} \Longrightarrow g(\mathbf{x}) + g(\mathbf{y}) \geq 1,$$

and let M be a subset of $[0, 1]^m \setminus \{\frac{1}{2}\mathbf{w}\}$ such that $\mathbf{x} \in M \iff \mathbf{w} - \mathbf{x} \notin M$ for all $\mathbf{x} \in [0, 1]^m \setminus \{\frac{1}{2}\mathbf{w}\}$.

The function $f_{IV1}(\cdot; g, k, M) : [0, 1]^{m+1} \to [0, 1]$ defined as

$$f_{IV1}(\mathbf{x}, x_{m+1}; g, k, M) := \begin{cases} 0, & \text{if } x_{m+1} < k, \\ 1, & \text{if } x_{m+1} > 1 - k, \\ 1/2, & \text{if } x_1 = x_2 = \cdots = x_{m+1} = 1/2, \\ g(\mathbf{x}), & \text{if } 1/2 < x_{m+1} \leq 1 - k \\ & \quad \text{or } (x_{m+1} = 1/2 \text{ and } \mathbf{x} \in M), \\ 1 - g(\mathbf{w} - \mathbf{x}), & \text{if } k \leq x_{m+1} < 1/2 \\ & \quad \text{or } (x_{m+1} = 1/2 \text{ and } \mathbf{w} - \mathbf{x} \in M), \end{cases} \quad (4.11)$$

is a VP-MDFF.

Note, here the condition (4.9), p. 100, to M needs not to be demanded.

The function (4.11), i.e. $f_{IV1}(\cdot, \cdot; g, k, M)$, becomes for $m = 1$ and $M = (\frac{1}{2}, 1]$ as follows:

$$f_{IV2}(x_1, x_2; g, k, (1/2, 1]) = \begin{cases} 0, & \text{if } x_2 < k, \\ 1, & \text{if } x_2 > 1 - k, \\ 1/2, & \text{if } x_1 = x_2 = 1/2, \\ g(x_1), & \text{if } 1/2 < x_2 \leq 1 - k \\ & \text{or } (x_2 = 1/2 \text{ and } x_1 > 1/2), \\ 1 - g(1 - x_1), & \text{if } k \leq x_2 < 1/2 \\ & \text{or } (x_2 = 1/2 \text{ and } x_1 < 1/2). \end{cases}$$

The following VP-MDFF is similar, but it differs for $x_1 \in \{0, 1\}$, i.e. in these cases it may get other function values.

Proposition 4.11 *Let g be a non-decreasing function defined from $[0, 1]$ to $[0, 1]$, such that*

$$g(y) + g(1 - y) \geq 1, \text{ for all } y \in [0, 1].$$

The function $f_{IV3}(\cdot; g, k) : [0, 1]^2 \to [0, 1]$ with $k \in \left(\frac{1}{3}, \frac{1}{2}\right]$ and defined as

$$f_{IV3}(\mathbf{x}; g, k) := \begin{cases} 0, & \text{if } x_1 = 0 \text{ or } (x_1 < 1 \text{ and } x_2 < k), \\ 1, & \text{if } x_1 = 1 \text{ or } (x_1 > 0 \text{ and } x_2 > 1 - k), \\ 1/2, & \text{if } x_1 = x_2 = 1/2, \\ g(x_1), & \text{if } (1/2 < x_2 \leq 1 - k \text{ and } 0 < x_1 < 1) \\ & \text{or } (x_2 = 1/2 \text{ and } 1/2 < x_1 < 1), \\ 1 - g(1 - x_1), & \text{if } (k \leq x_2 < 1/2 \text{ and } 0 < x_1 < 1) \\ & \text{or } (x_2 = 1/2 \text{ and } 0 < x_1 < 1/2), \end{cases}$$

is a VP-MDFF.

In the following proposition, we show how the g function in Proposition 4.11 should be defined so as to get the best lower bound that can be obtained from the corresponding VP-MDFF $f_{IV3}(\cdot; g, k)$ for the 2D-VPP.

Proposition 4.12 *Given an instance of the 2D-VPP, the best lower bound based on the function $f_{IV3}(\cdot; g, k)$ of Proposition 4.11 can be found with the following function $g : [0, 1] \to \{0, \frac{1}{2}, 1\}$ that depends on the parameters $s, t \in \mathbb{R}$ with $0 \leq s \leq \frac{1}{2}$ and $s \leq t \leq 1 - s$:*

$$g(x) := \begin{cases} 0, & \text{if } x < s, \\ 1/2, & \text{if } s \leq x \leq t, \\ 1, & \text{if } x > t. \end{cases}$$

4.3 Vector Packing Dual-Feasible Functions

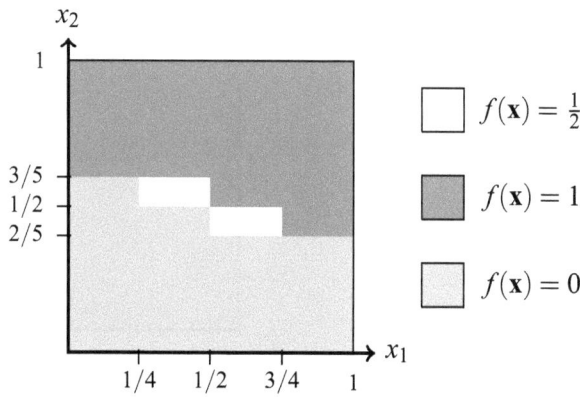

Fig. 4.6 Applying Proposition 4.11 to construct a VP-MDFF

Example 4.13 For a 2-dimensional example let us choose $s = 1/4, t = 1/2$ and $k = 2/5$. The obtained function is depicted in Fig. 4.6. □

4.3.3.5 Class V

Let $\mathbf{s}, \mathbf{t} \in (0, 1]^m$ be two constant vectors. Let u_1, u_2 be feasible (but not necessarily optimal) dual values for an instance of the m-dimensional vector packing problem with two items of sizes $\mathbf{s}$ and $\mathbf{t}$, each demanded at least once. The following proposition describes a new family of superadditive VP-DFF. Recall that these functions can be transformed into VP-MDFF by enforcing symmetry via Proposition 4.6, p. 100.

Proposition 4.13 *The function* $f_V(\cdot; \mathbf{s}, \mathbf{t}, u_1, u_2) : [0, 1]^m \to [0, 1]$ *is a superadditive VP-DFF:*

$$f_V(\mathbf{x}; \mathbf{s}, \mathbf{t}, u_1, u_2) := \max\{a_1 u_1 + a_2 u_2 | a_1, a_2 \in \mathbb{N}, a_1 \mathbf{s} + a_2 \mathbf{t} \leq \mathbf{x}\}.$$

Calculating the function requires solving an integer optimization problem for every argument $\mathbf{x}$. However, the complexity remains low, if the possible values $a_1, a_2 \in \mathbb{N}$ are bounded by a small constant.

Assume without loss of generality that

$$\max_{i \in \{1,\dots,m\}} s_i \leq \max_{i \in \{1,\dots,m\}} t_i.$$

Since one has only two items $\mathbf{s}, \mathbf{t} > \mathbf{o}$, the function value can be easily calculated by trying all possible numbers a_2, i.e. $a_2 \in \mathbb{N}$ and

$$a_2 \leq \min\{x_i/t_i : i \in \{1, \dots, m\}\},$$

Fig. 4.7 Class V function obtained from Example 4.14. The *bold line* corresponds to the segment $[(1, 0.4), (1, 1)]$ associated to value 1

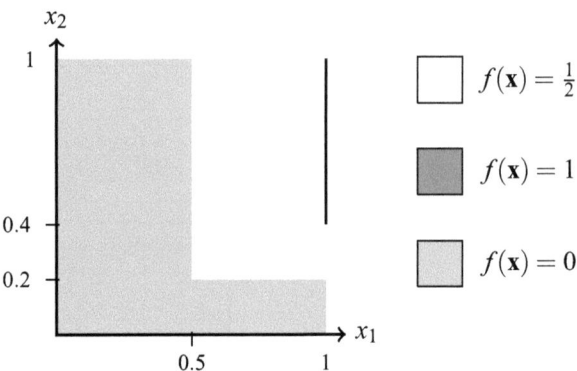

and setting

$$a_1 := \left\lfloor \min_{i \in \{1,\ldots,m\}} (x_i - a_2 \times t_i)/s_i \right\rfloor.$$

These calculations have the same effort as dynamic optimization with exactly two different items. Hence, the complexity to calculate the appropriate a_1 and a_2 is pseudo-polynomial.

Example 4.14 Consider the 2-dimensional example with $\mathbf{s} = (0.5, 0.2)^\top$ and $\mathbf{t} = (0.5, 0.6)^\top$. Pick the feasible dual values $u_1 = 0.5$ and $u_2 = 0.5$.

The optimization problem is

$$f(\mathbf{x}) = \max\{0.5a_1 + 0.5a_2 : 0.5a_1 + 0.5a_2 \leq x_1, 0.2a_1 + 0.6a_2 \leq x_2, a_1, a_2 \in \mathbb{N}\}.$$

The resulting superadditive but not maximal function is depicted in Fig. 4.7. It can be formulated as follows.

$$f(\mathbf{x}) = \begin{cases} 1, & \text{if } x_1 = 1 \text{ and } x_2 \geq 0.4 \\ 0, & \text{if } x_1 < 0.5 \text{ or } x_2 < 0.2 \\ 1/2, & \text{otherwise} \end{cases}$$

□

4.4 Orthogonal Packing

Here, we consider the m-dimensional bin-packing problem with and without rotation. Classical Cutting-Stock-DFF (CS-DFF) can be used to derive dual-feasible functions for this problem. Actually, most of the results in this section were initially

4.4 Orthogonal Packing

proposed without the general formalism of dual-feasible functions. Our presentation allows to gather different results from the literature under the same formalism.

We first define properly the m-dimensional Orthogonal Bin-Packing Problem, where one tries to pack m-dimensional rectangular bricks into large m-dimensional rectangular bricks. When rotation is allowed, since the bins are not always cubes, it is not possible to normalize the sizes in $[0, 1]^m$.

Problem 4.3 (m-Dimensional Orthogonal Bin-Packing Problem, m-OPP) An instance $D = (I; \mathbf{L}; \mathbf{w})$ of the m-OPP consists in a set $I = \{1, 2, \ldots, n\}$ of n items, whose sizes are given in the matrix $\mathbf{L} = (l_{11}, l_{12}, \ldots, l_{1m}; \ldots; l_{n1}, l_{n2}, \ldots, l_{nm}) \in \mathbb{N}^{n \times m}$ (with $\mathbf{l}_i$ being the ith row-vector of $\mathbf{L}$), and a rectangular brick described by its dimensions $\mathbf{w}$. The m-OPP consists in finding a partition of the set of items into a minimum number of subsets such that the items in each subset fit into a bin (i.e. the rectangular bricks fit into the boundary of the large rectangular brick, no two items overlap, and the edges of the items are parallel to the edges of the bin). If the rotation of the items is allowed, we have a m-OPP with rotation (m-OPP-R), otherwise, we have a m-OPP with fixed orientation (m-OPP-O).

The application of the concept of set-covering dual-feasible function to the orthogonal packing problem with and without rotation will be called m-OPP-R-DFF and m-OPP-O-DFF.

4.4.1 DFF for the Oriented Case (m-OPP-O-DFF)

One dual-feasible function can be applied to each dimension of an instance of m-OPP-O to obtain a lower bound. Similarly to the vector packing problem, the shape of each item i is described by a size vector $\mathbf{l}$. Thus, it is possible to design dual-feasible functions for this case that are independent of the data. Using the formalism introduced at the beginning of the section, the result can be written as follows.

Proposition 4.14 *Let $f_j : j = 1, \ldots, m$ be CS-DFF. The following function $g : \mathbb{R}^m \mapsto [0, 1]$ is a m-OPP-O-DFF.*

$$g(\mathbf{x}) := \prod_{j=1}^{m} f_j(x_j/w_j)$$

Example 4.15 Let $m = 2$, $\mathbf{w} = (10, 10)$ and consider four items of size $(6, 6)$. The trivial lower bound is equal to $\lceil 4 \times (36/100) \rceil = 2$. By taking $f_1 = f_{MT,0}(\cdot; \frac{1}{2})$ and $f_2 = f_{MT,0}(\cdot; \frac{1}{2})$, one obtains a bound equal to $\lceil 4 \times (1 \times 1) \rceil = 4$. □

If the identity function is used for $f_j, j = 1, \ldots, m$, the classical bound based on the surface/volume of the bins is obtained. No actual m-dimensional dual-feasible functions were derived in the literature (i.e. dual-feasible functions that would not consider the problem dimension by dimension). This can be explained by the fact that characterizing the set of feasible patterns is hard. Even verifying that a pattern is feasible is NP-complete.

4.4.2 DFF for the Case with Rotation (m-OPP-R-DFF)

The first result derives from a simple fact. For a set of CS-DFF $f_1,\ldots,f_m$, if a lower bound for the oriented case based on these functions is run for all possible orientations of the items, and if the minimum is recorded, a valid lower bound is obtained. Of course, the bound obtained would need an exponential time, since it would take $(m!)^n$ lower bounds to compute. Nevertheless a lower bound can be computed by considering the following relaxation: for each item i, keep the smallest image that it can have for its possible orientations. This leads to $n \times m!$ values to compute, which can lead to a practical method for lower bounding. Let S be the set of the $m!$ permutations $\sigma = \sigma(1),\ldots,\sigma(m)$ representing all possible orientations in m dimensions.

Proposition 4.15 *Let* $f_j : j = 1,\ldots,m$ *be CS-DFF. The following function* $\varphi_1 : \mathbb{R}^m \mapsto [0,1]$ *is a m-OPP-R-DFF.*

$$\varphi_1(\mathbf{x}) := \min_{\sigma \in S : x_{\sigma(j)} \leq w_j, j=1,\ldots,m} \left\{ \prod_{j=1}^{m} f_j(x_{\sigma(j)}/w_j) \right\}$$

A better m-OPP-R-DFF, that dominates the previous one if the f_j are increasing and superadditive, and if the container has equal size on each dimension, is now described.

Proposition 4.16 *Let* $f_j : j = 1\ldots,m$ *be m CS-DFF. If the instance of m-OPP-R is such that* $w_1 = w_2 = \ldots = w_m$ *then the following function* $\varphi_2 : \mathbb{R}^m \mapsto [0,1]$ *is a m-OPP-R-DFF.*

$$\varphi_2(\mathbf{x}) := \sum_{\sigma \in S} \left\{ \frac{\prod_{j=1}^{m} f_j(x_{\sigma(j)}/w_j)}{m!} \right\}$$

The result is not intuitive, but it becomes obvious when the following relaxation is considered. From a m-OPP-R instance, construct a m-OPP-O instance I' of size $m! \times n$ where each item is repeated once for each of its orientations. Clearly, the value of an optimal solution for this new problem cannot be more than $m!$ times the value of an optimal solution for the original m-OPP-R instance (see Fig. 4.8).

Example 4.16 Take $m = 2$, $\mathbf{w} = (10, 10)$ and four identical items $(8, 3)$, and choose the identity function for f_1 and $f_2 = f_{MT,0}(\cdot\,; 0.25)$, where the latter function was

4.5 Bin-Packing

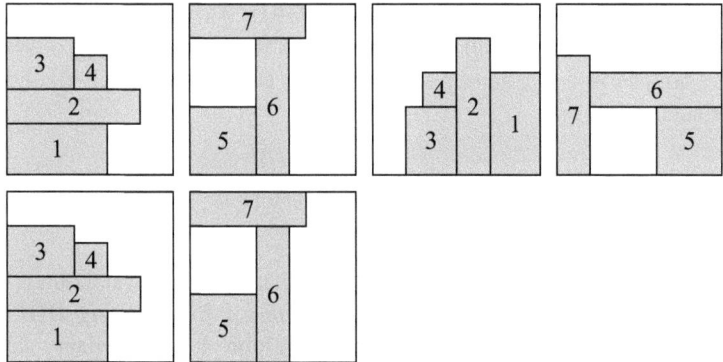

Fig. 4.8 An example of the relaxation of the OPP-R for $m = 2$. Note that for any solution for the OPP-R using z bins, there is always a solution of the duplicated OPP-O problem using $2z$ bins

defined in Formula (2.16), p. 48, as

$$f_{MT,0}(x; \lambda) := \begin{cases} 0, & \text{if } x < \lambda, \\ 1, & \text{if } x > 1 - \lambda, \\ x, & \text{otherwise.} \end{cases}$$

$\varphi_1((8, 3)) = \min\{0.8 \times f_{MT,0}(0.3; 0.25), 0.3 \times f_{MT,0}(0.8; 0.25)\} = \min\{0.8 \times 0.3, 0.3 \times 1\} = 0.24$. The bound obtained is equal to $\lceil 4 \times 24/100 \rceil = 1$.

$\varphi_2((8, 3)) = (0.8 \times f_{MT,0}(0.3; 0.25) + 0.3 \times f_{MT,0}(0.8; 0.25))/2 = (0.24 + 0.3)/2 = 0.27$. The bound obtained is equal to $\lceil 4 \times 0.27 \rceil = 2$. □

4.5 Bin-Packing

In this section, we present data-dependent dual-feasible functions designed for the bin-packing problem. For any item i, it is assumed that $0 < \ell_i \leq 1$. For our study, the only difference between bin-packing and cutting-stock is the fact that the number of items for each type is small and thus packing $\lfloor 1/\ell_i \rfloor$ instances of a given item i may not be allowed. This means that the set of valid patterns is smaller. In the data-dependent dual-feasible functions described in this section, the number of times an item is repeated in the instance can be taken into account.

Problem 4.4 (Bin-Packing Problem, BP) An instance $D = (I, \mathbf{l})$ of the OPP consists in a set $I = \{1, 2, \ldots, n\}$ of n items, whose sizes are given in the vector $\mathbf{l} \in [0, 1]^{n \times m}$. The Bin-Packing problem consists in finding a partition of the set of items into a minimum number of subsets such that the items in each subset fit into a bin, i.e. the sum of the sizes in each dimension does not exceed 1 for any subset.

In what follows, we use the classical 1-dimensional knapsack problem.

Problem 4.5 (Binary Knapsack Problem, KP–01) An instance $D = (J, \mathbf{l}, W, \alpha)$ of the binary knapsack problem is composed of a set J of items i, a size vector $\mathbf{l} \in [0, 1]_+^{|J|}$, a bin size $W \in \mathbb{R}_+$ and a profit vector $\alpha \in \mathbb{R}_+^{|J|}$. Formally, KP–01$(J, \mathbf{l}, W, \alpha)$ can be stated as follows.

$$\text{KP--01}(J, \mathbf{l}, W, \alpha) = \max \left\{ \sum_{i \in J} \alpha_i x_i : \sum_{i \in J} \ell_i x_i \leq W, x_i \in \{0, 1\}, \forall i \in J \right\}$$

Similarly to Cutting-Stock DFF (CS-DFF) that lead to lower bounds for the cutting-stock problem, one can define the notion of **Bin-Packing DDFF (BP-DDFF)** by specifying the subproblem of Definition 4.1 as a binary knapsack problem. The difference between the two classes of functions is that the polyhedral subproblem used in the definition is not the same (unbounded knapsack problem for the cutting-stock, and binary knapsack problem for bin-packing). Actually, any CS-DFF is a BP-DDFF, but the converse is generally not true.

Proposition 4.17 *Let $D = (I, \mathbf{l})$ be BP instance, $J \subseteq I$ a set of pairwise incompatible items ($\ell_i + \ell_j > 1, \forall i, j \in J$) and let $\alpha \in \mathbb{R}_+^n$. The following function $g_1 : I \to [0, 1]$ is a BP-DDFF defined for D.*

$$g_1(i) := \begin{cases} 1 & \text{if } i \in J \text{ and KP--01}(1, I \setminus J, \mathbf{l}, \alpha) = 0 \\ 1 - \text{KP--01}(1 - \ell_i, I \setminus J, \mathbf{l}, \alpha) / \\ \quad \text{KP--01}(1, I \setminus J, \mathbf{l}, \alpha) & \text{if } i \in J \text{ and KP--01}(1, I \setminus J, \mathbf{l}, \alpha) \neq 0 \\ 0 & \text{if } i \in I \setminus J \text{ and } \alpha_i = 0 \\ \alpha_i / \text{KP--01}(1, I \setminus J, \mathbf{l}, \alpha) & \text{if } i \in I \setminus J \text{ and } \alpha_i \neq 0 \end{cases}$$

The percentage of the bin taken by each small item i is equal to α_i, and the sizes of the large items are computed by solving the knapsack problem described above. Note that in some degenerate cases, the image of an item in J may be smaller than the image of an item in $J \setminus I$.

Example 4.17 Consider a BP-instance $(I, \mathbf{l})$, with $I = \{1, \ldots, 10\}$ and

$$\mathbf{l} = (0.1, 0.1, 0.1, 0.2, 0.3, 0.3, 0.3, 0.3, 0.6, 0.6).$$

Choose $J = \{9, 10\}$ and $\alpha = (0, 0, 0, 1, 1, 1, 1, 1, 6, 6)$.

$$\text{KP--01}(\{1, \ldots, 8\}, \mathbf{l}, \alpha) = 3.$$
$$g_1(1) = g_1(2) = g_1(3) = 0,$$
$$g_1(4) = g_1(5) = g_1(6) = g_1(7) = g_1(8) = 1/3,$$
$$g_1(9) = g_1(10) = 1 - 1/3 = 2/3.$$

Note that the obtained values cannot be computed by the means of a dual-feasible function, since $g_1(4) = 1/3$, which would not be possible because $\ell_4 = 0.2$ and for any dual-feasible function g, $g(0.2) \leq 1/5$ (otherwise the dual-feasible function condition would not hold for five items of size 0.2). □

The knapsack problems involved are NP-hard in the general case. However, they can be solved in pseudo-polynomial time using dynamic programming. When the size of the bin is large, it may entail a large computing time. In this case, the set of parameters α should be chosen in a way to re-enable the resolution of the knapsack problem in a polynomial time (for example $\alpha_i = 1, \forall i \in J$).

4.6 Bin-Packing Problem with Conflicts

We now describe data-dependent dual-feasible functions for the bin-packing with conflicts. In this problem, some pairs of items cannot be packed in the same bin. A natural way of modelling these constraints is to use a graph. Note that we consider the generic case without specifying the geometric constraint applied to the problem. Similarly to BP-DDFF, we will name BPC-DDFF the DDFF designed for the bin-packing problem with conflicts. Only data-dependent functions are useful here, since it makes no sense to define a function that would be valid for any graph. Actually the class of BPC-DFF would be equivalent to the class of CS-DFF, because the instances with a complete compatibility graph have to be considered.

Problem 4.6 (Bin-Packing Problem with Conflicts, BPC) An instance $D = (I; \mathbf{l}; E)$ of the m-OPP consists in a set $I = \{1, 2, \ldots, n\}$ of n items, whose sizes are given in the vector $\mathbf{l} \in [0, 1]^n$, and $E \subseteq I \times I$ a set of compatibility edges. The Bin-Packing with conflicts consists in finding a partition of the set of items into a minimum number of subsets such that the items in each subset fit into a bin, i.e. the sum of the sizes in each dimension does not exceed 1 for any subset, and for any two items i, j in the same partition $(i, j) \in E$.

Both techniques described below are based on graph-theoretical concepts, namely graph triangulation, and tree-decomposition, which are defined next.

4.6.1 BPC-DDFF Based on a Knapsack Subproblem

The first BPC-DDFF is a generalization of function g_1 defined in Proposition 4.17 for BP. It has to solve knapsack problems with conflicts instead of a classical knapsack problem.

Problem 4.7 (Binary Knapsack Problem with Conflicts, KPC) An instance $D = (J, \mathbf{l}, W, \alpha, E)$ of Binary Knapsack Problem with Conflicts consists of an item set $J = \{1, \ldots, n\}$, whose sizes are given in the vector $\mathbf{l} \in [0, 1]^n$, a profit vector $\alpha \in \mathbb{R}^n$, and E a set of compatibility edges. Formally, $KPC(J, \mathbf{l}, W, \alpha, E)$ can be stated as follows.

$$KPC(J, \mathbf{l}, \alpha, E) = \max \left\{ \sum_{i \in J} \alpha_i x_i : \sum_{i \in J} \ell_i x_i \leq W, x_i + x_j \leq 1, \forall (i,j) \notin E, x_i \in \{0,1\}, \forall i \in I \right\}$$

In the following proposition, and in the remainder of the book, for a vertex i, $N(i) = \{i\} \cup \{j : (i,j) \in E\}$ is the neighbourhood of i.

Proposition 4.18 *Let $D = (I; \mathbf{l}; E)$ be a BPC instance, J a set of pairwise incompatible items, and $\{\alpha_i \in \mathbb{R}_+, i \in I \setminus J\}$ a list of coefficients. Function $g_1(\cdot; J, \alpha) : I \to [0,1]$ is a BPC-DDFF.*

$$g_1(i) := \begin{cases} 1 - KPC(1 - \ell_i, N(i), \mathbf{1}, \alpha)/KPC(1, I \setminus J, \mathbf{1}, \alpha) & \text{if } i \in J \\ \alpha_i/KPC(1, I \setminus J, \mathbf{1}, \alpha) & \text{if } i \in I \setminus J \end{cases}$$

Example 4.18 Consider an instance $D = (I; \mathbf{l}; E)$ where $I = \{1, 2, 3, 4\}$,

$$\mathbf{l} = (0.7, 0.6, 0.3, 0.2)$$

and

$$E = \{(2,3), (2,4), (3,4)\}.$$

Let $J = \{1, 2\}$, and $\alpha_3 = 1, \alpha_4 = 1$.

$$KPC(W, J, \mathbf{1}, \alpha) = 2.$$
$$g_1(3; J, \alpha) = g_1(4; J, \alpha) = 1/2,$$
$$g_1(1; J, \alpha) = 1 - 0 = 1,$$
$$g_1(2; J, \alpha) = 1 - 1/2 = 1/2.$$

The obtained bound is equal to $\lceil 1 + 1/2 + 1/2 + 1/2 \rceil = 3$.

Any bound based on a cutting-stock dual-feasible function would have been at most two, because the optimum of the related cutting-stock problem is 2 according to the partition $I = \{1, 3\} \cup \{2, 4\}$. Graph colouring would also yield the bound 2 only, since two colours are enough for colouring the complementary graph according to $\{1\}, \{2, 3, 4\}$. Both bounds are less than the obtained bound 3. □

To the best of our knowledge, no dynamic programming scheme exists for the disjunctive knapsack problem with general graphs. When a conflict graph G is considered, only cliques of G can be solutions of the knapsack problem with conflicts. Thus a (possibly not practically tractable) solution for the latter is to

4.6 Bin-Packing Problem with Conflicts

compute all maximal cliques of the conflict graph, and then to solve for each clique the associated knapsack problem. The maximum value obtained for all cliques is the optimal value for the knapsack problem with conflicts. This solution is tractable only if the cliques are in small number, and they can be computed with a small complexity. Neither of the two conditions are fulfilled when a random graph is considered. For this method to be tractable, the problem can be relaxed by adding edges to the compatibility graph in such a way that it becomes *triangulated*. A graph G is *triangulated* if for every cycle of length $k > 3$, there is a chord joining two non-consecutive vertices. Any triangulated graph G has at most n maximal cliques. In addition, they can be computed in linear time. Finding the minimum set of edges to add in order to obtain a triangulated graph is a NP-hard problem, so a heuristic should be used.

4.6.2 A BPC-DDFF Based on Graph Decomposition

Suppose the set I of items can be decomposed into two sets I_1 and I_2 of pairwise incompatible items. In this case, two different dual-feasible functions f and g can be applied to I_1 and I_2, since the instance can be decomposed into two distinct sub-instances. Now, if there is a third set I_3 where each item is compatible with some items of I_1 and I_2, each item of I_3 will be packed either with items of I_1, items of I_2, or neither of these items, but not both. This leads to the following BPC-DDFF, which depends on two CS-DFF f and g.

Proposition 4.19 *Let $D = (I; 1; E)$ be an instance of BPC, and let also (I_1, I_2, I_3) be a partition of I such that $E \cap \{(i_1, i_2) : i_1 \in I_1, i_2 \in I_2\} = \emptyset$. Let also f and g be two CS-DFF. Function $h(\cdot; f, g, I_1, I_2) : I \to [0, 1]$ defined as follows is a BPC-DDFF.*

$$i \mapsto \begin{cases} f(\ell_i) & \text{if } i \in I_1 \\ g(\ell_i) & \text{if } i \in I_2 \\ \min\{f(\ell_i), g(\ell_i)\} & \text{otherwise} \end{cases}$$

This technique can be generalized by decomposing the graph into different intersecting subsets. Function h_2 is based on the concept of *tree-decomposition*, which captures the possible associations of items.

Definition 4.5 A *tree-decomposition* of $G = (I, E)$ is a pair (X, T), where $T = (V, A)$ is a tree with node set V and edge set A, and $X = \{X_v : v \in V\}$, is a family of subsets of I such that:

1. $\cup_{v \in V} X_v = I$
2. $\forall [v, w] \in A$, there is a X_v, $v \in V$ with $v \in X_v$ and $w \in X_v$
3. $\forall i, j, k \in V$, if j is on the path from i to k in T, then $X_i \cap X_k \subseteq X_j$

Let (X, T) be a tree-decomposition of G. The basic idea of h_2 is to assign a given DFF f_s to each subset $s \in X$. Let F be a list of CS-DFF $f_1, \ldots, f_{|S|}$, one for each node of the tree decomposition. Recall that S is the set of all permutations representing the possible orientations. For each vertex i in the graph we define S_i the set of nodes of the tree decomposition containing i. Clearly there is always a set of functions $f_1, \ldots, f_{|S|}$ that allows to dominate the application of a single DFF (e.g. $f_1 = f_2 = \ldots = f_{|S|}$).

Proposition 4.20 *Let $D = (I; \mathbf{l}; E)$ be an instance of BPC, and (X, T) be a tree-decomposition of G. Let also $F = f_1, \ldots, f_{|S|}$ be a list of CS-DFF, and for each vertex $i \in I$, let S_i be the set of nodes of the tree decomposition containing i. The following function $h_2(\cdot; F, (X, T)) : I \to [0, 1]$ is a BPC-DDFF.*

$$i \mapsto \min_{s \in S_i} \{f_s(\ell_i)\}$$

An issue is to choose a suitable set of functions to be applied to the nodes of the tree decomposition.

Example 4.19 Let $I = \{1, \ldots, 7\}$,

$$\mathbf{l} = (0.2, 0.2, 0.3, 0.4, 0.8, 0.5, 0.5),$$

and

$$E = \{(1, 2), (1, 3), (1, 4), (2, 3), (2, 4), (3, 4), (3, 5), (4, 5), (4, 6), (5, 7), (6, 7)\}$$

(see Fig. 4.9).

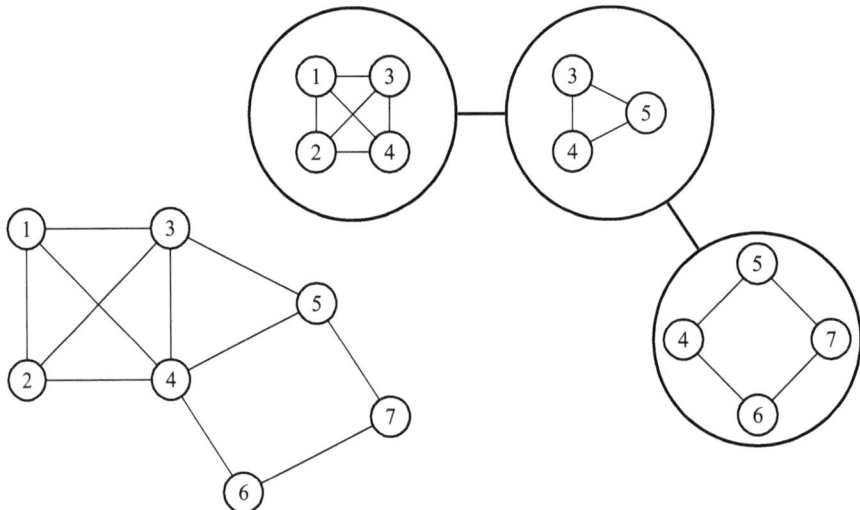

Fig. 4.9 Compatibility graph for Example 4.19 and a possible tree-decomposition for this graph

Table 4.1 Values of the functions obtained for the different nodes of the tree decomposition

Function	Items						
	1	2	3	4	5	6	7
f_1	0.2	0.2	0.3	0.4			
f_2			0.3	0.4	1		
f_3				0.4	1	0.5	0.5
min	0.2	0.2	0.3	0.4	1	0.5	0.5

Let (X, T) be a tree decomposition of (I, E) such that

$$V_1 = \{1, 2, 3, 4\}, \ V_2 = \{3, 4, 5\}, \text{ and } V_3 = \{4, 5, 6, 7\}.$$

Let $f_1 = id$ and $f_2 = f_3 = f_{MT,0}(\cdot; 0.25)$. The values obtained are reported in Table 4.1.

The obtained bound is equal to

$$\lceil 0.2 + 0.2 + 0.3 + 0.4 + 1 + 0.5 + 0.5 \rceil = 4.$$

Note that this bound is better than :

- any that would be designed for bin-packing only, since the optimal solution is 3 bins: $\{1, 5\}, \{2, 3, 4\}, \{6, 7\}$;
- any graph-colouring bound, since there is a proper colouring of the complementary graph using 3 colours only: $\{1, 2, 3, 4\}, \{5\}, \{6, 7\}$.

□

4.7 Related Literature

Dual-feasible functions for vector packing problems were proposed by Alves et al. (2014). Applying dual-feasible functions to orthogonal packing problem has been done for the first time by Fekete and Schepers (2004). It has also been done implicitly by Boschetti and Mingozzi (2003). In Carlier et al. (2007), the functions implicitly used by Boschetti and Mingozzi (2003) were described and slightly improved. Caprara et al. (2005) show that applying a dual-feasible function on each dimension of an 2-dimensional orthogonal packing problem often leads to a bound of excellent quality. The notion of data-dependent functions was proposed by Carlier et al. (2007). Clautiaux et al. (2007) show that dual-feasible functions can be used to produce bounds for the orthogonal packing problem with rotation. Dual-feasible functions for the case with conflicts were introduced by Khanafer et al. (2010). Everything the reader needs to know about treewidth and graph triangulation is respectively available in Rose et al. (1976) and Robertson and Seymour (1986).

4.8 Exercises

1. For each of the two following functions f_1 and f_2 defined from $[0, 1]^2$ to $[0, 1]$, indicate whether or not the function is a VP-DFF, and if it is a VP-MDFF. If the function is a VP-DFF and not a VP-MDFF, propose a VP-MDFF that dominates it.

$$f_1(\mathbf{x}) := \begin{cases} 1 & \text{if } x_1 > 1/2 \text{ and } x_2 > 1/2 \\ 0 & \text{otherwise} \end{cases}$$

$$f_2(\mathbf{x}) := \begin{cases} 1 & \text{if } x_1 > 1/2 \text{ or } x_2 > 1/2 \\ 0 & \text{otherwise} \end{cases}$$

2. Is function f_3 a 2-OPP-O-DFF?

$$f_3(\mathbf{x}) := \begin{cases} 1 - \lfloor \frac{w_2 - x_2}{4} \rfloor, & \text{if } x_1 > 2w_1/3 \text{ and } x_2 > w_2/2, \\ 1/2, & \text{if } x_1 > 2w_1/3 \text{ and } x_2 = w_2/2, \\ \lfloor \frac{x_2}{4} \rfloor, & \text{if } x_1 > 2w_1/3 \text{ and } x_2 < w_2/2, \\ \left(1 - \lfloor \frac{w_2 - x_2}{4} \rfloor \right), & \text{if } 2w_1/3 \geq x_1 \geq w_1/3 \text{ and } x_2 > w_2/2, \\ x_1/2 & \text{if } 2w_1/3 \geq x_1 \geq w_1/3 \text{ and } x_2 = w_2/2, \\ x_1 \times \lfloor \frac{x_2}{4} \rfloor & \text{if } 2w_1/3 \geq x_1 \geq w_1/3 \text{ and } x_2 < w_2/2, \\ 0, & \text{if } x_1 < w_1/3. \end{cases}$$

3. Describe the 2-OPP-O-DFF obtained using Proposition 4.14 with functions $f_{CCM,1}(\cdot; 4)$ and $f_{CCM,1}(\cdot; 2)$. Apply it to the following instance: $\mathbf{W} = (10, 10)$, $I = \{1, \ldots, 6\}$ and

$$\mathbf{L} = ((5, 4), (5, 4), (6, 4), (6, 4), (8, 8), (2, 2)).$$

4. Compute the 2-OPP-R-DFF obtained using Proposition 4.15 and functions $f_{MT,0}(\cdot; 0.3)$ and $f_{CCM,1}(\cdot; 3)$. Apply it to instance $\mathbf{W} = (10, 10)$, $I = \{1, \ldots, 6\}$, and

$$\mathbf{L} = ((7, 7), (8, 8), (4, 7), (4, 7), (7, 3), (7, 3)).$$

Do the same using Proposition 4.16.

4.8 Exercises

5. Consider the 2-OPP-R. Let f and g be two CS-MDFF. For $w > h$, let

$$F^{f,g}(w,h) := \lfloor w/h \rfloor f(h/W) \times g(h/H) + F^{f,g}(h, w \bmod h);$$
$$F^{f,g}(w, 0) := 0.$$

1. Prove that if $w, h \in \mathbb{N}, w \geq h$, and if f and g can be computed in finite time, then $F^{f,g}(w,h)$ can be computed in finite time too.
2. Prove that $\varphi_3 : i \mapsto F^{f,g}(w_i, h_i)$ is a 2-OPP-R-DFF.
3. Show that if two CS-MDFF f and g are used, if the bin is a square, and if each item i is such that $\exists k, w_i = k \times h_i$, then

$$\varphi_3(i) \leq \varphi_1(i) \leq \varphi_2(i),$$

$\forall i \in I$ (with φ_1, φ_2 from Propositions 4.15 and 4.16).

4. Show that if two CS-MDFF f and g are used, if the bin is a square, then

$$\varphi_3(i) \leq \varphi_2(i)$$

$\forall i \in I$ (with φ_2 from Proposition 4.16).

6. Let us introduce a new variant of the bin-packing problem: the multi-mode bin-packing. In this problem, each item can have several "modes", i.e. it has a finite number of possible sizes. The choice of the mode is a part of the optimization. Propose a MMBP-DFF.

7. Indicate whether the mappings f_1, f_2 and f_3 are data-dependent dual-feasible functions for the instance $\mathbf{l} = (0.2, 0.2, 0.3, 0.3, 0.3, 0.6, 0.7)$.

i	1	2	3	4	5	6	7
l_i	0.2	0.2	0.3	0.3	0.3	0.6	0.7
$f_1(i)$	0.3	0.3	0.2	0.2	0.2	0.4	0.7
$f_2(i)$	0.2	0.3	0.25	0.25	0.25	0.5	0.7
$f_3(i)$	0.2	0.3	0.2	0.25	0.25	0.5	0.7

8. Propose a strictly better data-dependent dual-feasible function than function f for the instance $W = 1, I = \{1, \ldots, 7\}, \mathbf{l} = \{0.3, 0.3, 0.6, 0.6, 0.8, 0.8, 0.9\}$.

i	1	2	3	4	5	6	7
l_i	0.3	0.3	0.6	0.6	0.8	0.8	0.9
$f(i)$	0.33	0.33	0.5	0.5	1	1	1

9. Show that increasing the size of an item may decrease the value of the bound obtained using the BP-DDFF defined in Proposition 4.17. Can it happen with a CS-MDFF?

10. Apply Proposition 4.18 to obtain a BPC-DDFF for the following instance of BPC: $W = 10, I = \{1, \ldots, 14\}$,

$$\mathbf{l} = (0.8, 0.8, 0.8, 0.8, 0.2, 0.7, 0.3, 0.6, 0.3, 0.6, 0.3, 0.5, 0.5, 0.4)$$

are the sizes and

$$E = I \times I \setminus \{(8, 14), (9, 14), (10, 14), (11, 14), (12, 13), (12, 14), (13, 14)\}$$

the arcs. Choose your coefficients α_i in such a way that the lower bound produced is optimal.

11. Apply Proposition 4.20 to obtain a BPC-DDFF for the following instance of BPC: $W = 1, I = \{1, \ldots, 12\}$,

$$\mathbf{l} = (0.4, 0.3, 0.3, 0.3, 0.3, 0.3, 0.7, 0.7, 0.7, 0.7, 0.3, 0.3),$$

and the compatibility edge set is

$$E = \{(1, 2), (1, 3), (1, 4), (2, 3), (2, 4), (3, 4), (3, 5), (3, 8), (2, 10),$$
$$(1, 7), (1, 9), (9, 10), (5, 6)(6, 11), (11, 12), (4, 12), (7, 8)\}$$

(see below). Choose your CS-DFF in such a way that the lower bound produced is optimal.

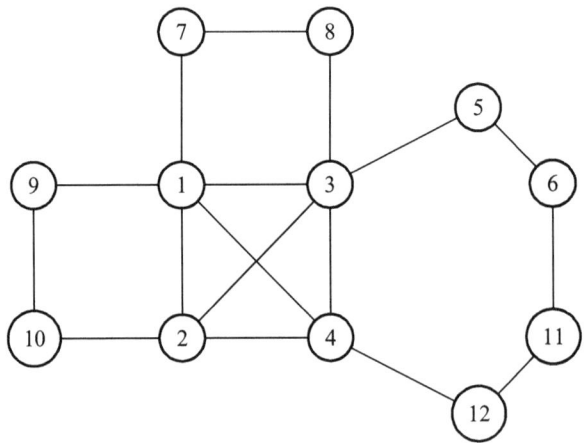

Hint: use a tree-decomposition with four nodes.

4.8 Exercises

12. Consider an instance of BPC such that $\sum_{i \in I} \ell_i \leq 1$, and $G = (I, E)$ is co-interval. Propose a BPC-DDFF for this problem that always leads to an optimal solution.

Hint: the problem is a well-known polynomial case of one of the most famous hard combinatorial problems.

Chapter 5
Other Applications in General Integer Programming

In this chapter, we briefly review an alternative application of dual-feasible functions in general integer programming. We explore these functions in particular to derive valid inequalities for integer programs. Since the notion of superadditivity is essential for this purpose, we start by reviewing superadditivity in the scope of valid inequalities. Different examples are provided with alternative families of dual-feasible functions. We discuss also the difference between the valid inequalities derived by dual-feasible functions and the well-known Chvátal-Gomory cuts.

5.1 Superadditive Functions in Integer Programming

When good dual-feasible functions are sought, they are frequently characterized by superadditivity and monotonicity. For the sake of clarity, these properties are briefly recalled in the sequel for general domains. Given $X \subseteq \mathbb{R}^m$, a function $F : X \to \mathbb{R}$ is *superadditive* if for all $\mathbf{x}, \mathbf{y} \in X$ with $\mathbf{x} + \mathbf{y} \in X$, it holds that

$$F(\mathbf{x}) + F(\mathbf{y}) \leq F(\mathbf{x} + \mathbf{y}).$$

The function $F : X \to \mathbb{R}$ is *nondecreasing* if for all $\mathbf{x}, \mathbf{y} \in X$, one has that

$$\mathbf{x} \leq \mathbf{y} \implies F(\mathbf{x}) \leq F(\mathbf{y}).$$

Analogously to the duality theory of linear optimization (without integer constraints), there is also a *strong duality theorem* in the discrete counterpart. Here superadditive and nondecreasing functions are essential. Given $\mathbf{A} \in \mathbb{Q}^{m \times n}$, $\mathbf{b} \in \mathbb{Q}^m$ and $\mathbf{c} \in \mathbb{Q}^n$ and an integer linear optimization problem

$$\max \ \mathbf{c}^T\mathbf{x} \qquad (5.1)$$
$$\text{s. to } \ \mathbf{Ax} \leq \mathbf{b} \qquad (5.2)$$
$$\mathbf{x} \in \mathbb{Z}_+^n, \qquad (5.3)$$

the dual consists in determining a nondecreasing and superadditive function $F : \mathbb{R}^m \to \mathbb{R}$ such that

$$\min \ F(\mathbf{b}) \qquad (5.4)$$
$$\text{s. to } \ F(\mathbf{o}) = 0 \qquad (5.5)$$
$$F(\mathbf{a}^j) \geq c_j \text{ for } j = 1, \ldots, n, \qquad (5.6)$$

where $\mathbf{a}^j$ is the j-th column of the matrix $\mathbf{A}$, and $\mathbf{o}$ denotes as usual the zero vector. As mentioned before, the relationship between the two problems is given by the *strong duality theorem*. If the problem (5.1)–(5.3) is solvable, i.e. a feasible $\mathbf{x}$ exists and the objective function value is bounded from above, then the optimal objective function values of the primal and the dual problem are equal. If $F(\mathbf{b})$ is unbounded from below ($-\infty$), then there is no $\mathbf{x} \in \mathbb{Z}_+^n$ fulfilling (5.2), and if (5.1)–(5.3) yields an unbounded objective function value ($+\infty$) then F does not exist. Furthermore, if $\hat{\mathbf{x}}$ is an optimal solution of (5.1)–(5.3) then any optimal F in the dual problem necessarily obeys the equation

$$F(\mathbf{Ax}) = \mathbf{c}^T\mathbf{x} = F(\mathbf{b}) - F(\mathbf{b} - \mathbf{Ax})$$

for all $\mathbf{x} \in \mathbb{Z}_+^n$ with $\mathbf{x} \leq \hat{\mathbf{x}}$.

Since the problem (5.1)–(5.3) is NP-hard, solving it exactly or finding an optimal superadditive and nondecreasing function $F : \mathbb{R}^m \to \mathbb{R}$ according to the above conditions (5.4)–(5.6) may be very difficult. However, contributing to the resolution of (5.1)–(5.3) is possible by deriving valid inequalities using a superadditive and nondecreasing function $F : \mathbb{R}^m \to \mathbb{R}$ fulfilling the constraints (5.5). These functions lead to the following valid inequalities:

$$\sum_{j=1}^n F(\mathbf{a}^j) \times x_j \leq F(\mathbf{b}).$$

5.2 Valid Inequalities for Integer Programs

General dual-feasible functions can be used not only to compute fast lower bounds, but also to generate valid inequalities for integer problems defined over sets of the kind $\{\mathbf{x} \in \mathbb{Z}_+^n : \mathbf{Ax} \leq \mathbf{b}\}$, as stated formally in the sequel.

If $r > 0$, then dividing the inequality (5.7) by r and applying a maximal general dual-feasible function $f : \mathbb{R} \to \mathbb{R}$ with $f(1) = 1$ leads to the valid inequality

$$\sum_{j=1}^{n} f\left(\frac{d_j}{r}\right) \times x_j \leq 1. \tag{5.9}$$

In some situations it may happen that the inequality (5.9) is stronger than (5.8), but not if $0 < r < 1$, as the right-hand sides show immediately. The following example demonstrates the strength of a maximal general dual-feasible function for the construction of valid inequalities, even in the case $r = 1$.

Example 5.1 We use the function (3.7), p. 65, with the parameter $b := 1$, hence

$$f(d) = \begin{cases} \lfloor 2d \rfloor, & \text{if } d < 1/2, \\ 1/2, & \text{if } d = 1/2, \\ \lceil 2d \rceil - 1, & \text{if } d > 1/2. \end{cases} \tag{5.10}$$

Let be given the inequality

$$1.6x_1 - 0.4x_2 \leq 1.$$

Of course, if there would be no negative coefficient then a coefficient $d_j > 1$ would immediately imply $x_j = 0$ for that j. The Chvátal-Gomory cut (5.8) transforms the given inequality to $x_1 - x_2 \leq 1$, but the function (5.10) used in (5.9) leads to the stronger inequality $3x_1 - x_2 \leq 1$. This inequality would be obtained by the Chvátal-Gomory procedure only after multiplying the given inequality by a suitable number like 1.9. □

Example 5.2 The dual-feasible function (2.10) is used for deriving valid inequalities. It was defined as

$$f_{FS,1}(x; k) := \begin{cases} x, & \text{if } (k+1) \times x \in \mathbb{N}, \\ \lfloor (k+1) \times x \rfloor / k, & \text{otherwise}. \end{cases}$$

and is again illustrated in Fig. 5.1. Next, we show the result of applying the function $f_{FS,1}$ with parameters $k \in \{1, 2, 3\}$ to a given knapsack inequality (after dividing it by the right-hand side):

$$9x_1 + 7x_2 + 6x_3 + 4x_4 + 2x_5 \leq 12$$

a_i/b	0.75	0.58	0.5	0.33	0.17		
$k = 1$	$1x_1$	$+ 1x_2$	$+ \frac{1}{2}x_3$			$\leq$	1
$k = 2$	$1x_1$	$+ \frac{1}{2}x_2$	$+ \frac{1}{2}x_3$	$+ \frac{1}{3}x_4$		$\leq$	1
$k = 3$	$\frac{3}{4}x_1$	$+ \frac{2}{3}x_2$	$+ \frac{1}{2}x_3$	$+ \frac{1}{3}x_4$		$\leq$	1

□

Proposition 5.1 *If f is a maximal general dual-feasible function (and hence a superadditive function with $f(0) = 0$) and $S = \{\mathbf{x} \in \mathbb{Z}_+^n : \sum_{j=1}^n a_{ij}x_j \leq b_i, i = 1, \ldots, m\}$, then for any i, $\sum_{j=1}^n f(a_{ij})x_j \leq f(b_i)$ is a valid inequality for S.*

In Chaps. 2 and 3 it turned out that (general) dual feasible functions should be superadditive to get good lower bounds, because otherwise another dominating function exists. In the context of valid inequalities for integer problems, the same applies. Any valid inequality for S can be obtained either through a superadditive function or it is dominated by an inequality that can be computed in this way. Cuts generated by a superadditive function are commonly referred to as *superadditive inequalities*. Among these cuts, those that are not dominated by any other valid inequality are called maximal. This applies particularly to the facets of the integer hull of S. Maximal valid inequalities are necessarily superadditive. The same properties characterize the dominant families of dual-feasible functions.

Any maximal inequality for S can be obtained through the Gomory procedure based on recursive linear combinations and rounding of other inequalities for S. However, in order to get these maximal cuts, it might be necessary to use a very long recursion. Other authors assumed that other superadditive functions, eventually more complex ones, might generate these maximal cuts using shorter recursions, demonstrating the relevance of research on dual-feasible functions as tools to compute valid inequalities for integer programs. Other works propose alternative characterizations of the integer hull of S in terms of a finite set of superadditive inequalities, but the cardinality of this set may be very large, such that many cuts are required.

5.3 Examples

In this section, we show through different examples how to apply several generalized dual-feasible functions to derive valid inequalities, which may be better than the well-known Chvátal-Gomory cuts.

Given the inequality system $\mathbf{Ax} \leq \mathbf{b}$, $\mathbf{x} \in \mathbb{N}^n$, any nonnegative linear combination of the inequalities may be used to derive cuts. Choosing $\mathbf{u} \geq \mathbf{o}$ yields $\mathbf{u}^\top \mathbf{Ax} \leq \mathbf{u}^\top \mathbf{b}$, hence the following scalar inequality

$$\mathbf{d}^\top \mathbf{x} \leq r, \tag{5.7}$$

and finally the Chvátal-Gomory cut

$$\sum_{j=1}^n \lfloor d_j \rfloor \times x_j \leq \lfloor r \rfloor. \tag{5.8}$$

5.3 Examples

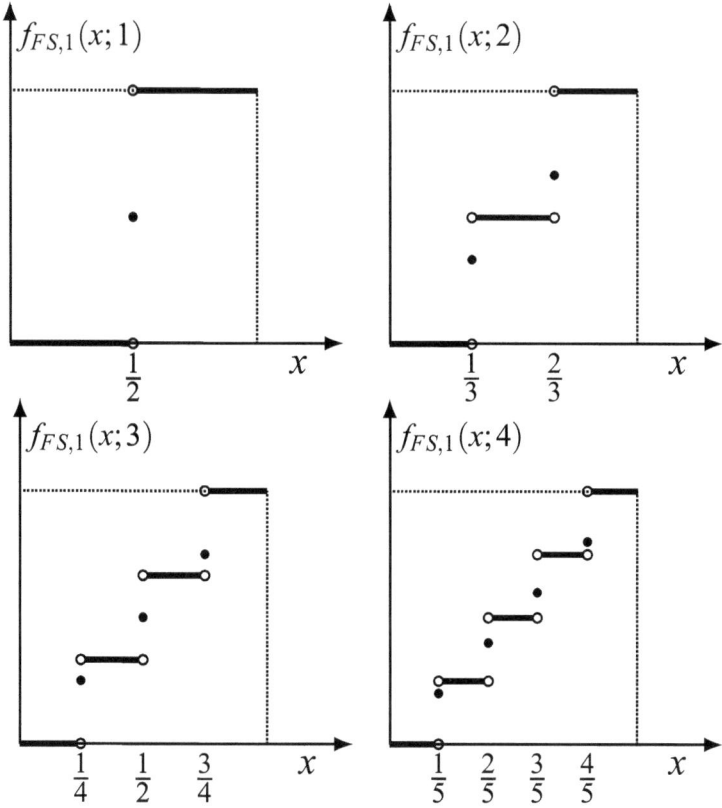

Fig. 5.1 MDFF $f_{FS,1}(\cdot;k)$ for $k \in \{1,\ldots,4\}$

Given multiple knapsack constraints like in the vector packing problem, one may use a VP-MDFF to get cuts.

Example 5.3 Neglecting the integrality constraints in

$$8x_1 + 5x_2 + 5x_3 \leq 10$$
$$5x_1 + 8x_2 + 5x_3 \leq 10$$
$$x_1, x_2, x_3 \in \{0, 1\}$$

would allow the fractional solution $\mathbf{x} = (0.5, 0.8, 0.2)^\top$, because

$$8 \times 0.5 + 5 \times 0.8 + 5 \times 0.2 = 4 + 4 + 1 = 9 \leq 10$$

and

$$5 \times 0.5 + 8 \times 0.8 + 5 \times 0.2 = 2.5 + 6.4 + 1 = 9.9 \leq 10.$$

However, taking e.g. the VP-MDFF (4.9), p. 106, with the parameter $\mathbf{u} := (\frac{1}{2}, \frac{1}{2})^\top$, which considers all constraints simultaneously, yields the valid inequality

$$x_1 + x_2 + 0.5x_3 \leq 1,$$

which is violated by the fractional solution, since

$$0.5 + 0.8 + 0.5 \times 0.2 = 1.4 > 1.$$

□

The following example illustrates the generation of valid inequalities with dual-feasible functions obtained from bin packing problems with conflicts.

Example 5.4 Consider the following system of inequalities:

$$4x_1 + 3x_2 + 3x_3 \leq 10$$
$$x_2 + x_3 \leq 1$$
$$x_1, x_2, x_3 \in \{0, 1\}$$

The first inequality does not restrict anything. However, a BPC-DFF yields the valid inequality

$$0.9x_1 + 0.1x_2 + 0.1x_3 \leq 1.$$

The DFF used is the one defined in Proposition 4.17, with $J = \{2, 3\}$, $\alpha_2 = 0.1$ and $\alpha_3 = 0.1$.

□

5.4 Related Literature

In Nemhauser and Wolsey (1998), the relationship between the integer linear optimization problem (5.1)–(5.3) and its dual (5.4)–(5.6) and a revision of superadditive valid inequalities are provided, together with the basic function underlying the Chvátal-Gomory procedure. Previous results on the (explicit or implicit) use of dual-feasible functions to generate valid inequalities for integer programs were reported by Vanderbeck (2000), Alves (2005), Rietz et al (2014), Letchford and Lodi (2002), and Dash and Günlük (2006).

5.5 Exercises

1. Let be given the feasible region $\sum_{j=1}^{n} a_{ij}x_j \leq b_i$, $i = 1, \ldots, m$, $\mathbf{x} \in \mathbb{N}^n$ for an integer linear optimization problem. Which of the following assertions are true? Justify your answer.

   (a) $\sum_{i=1}^{n} f(a_{ij})x_j \leq f(b_i)$ is a valid inequality for every general DFF $f : \mathbb{R} \to \mathbb{R}$.

   (b) $\sum_{i=1}^{n} f(a_{ij})x_j \leq f(b_i)$ is a valid inequality for every superadditive function $f : \mathbb{R} \to \mathbb{R}$.

2. Consider the set of all ordered triplets $(x_1, x_2, x_3) \in \mathbb{Z}_+^3$, for which

$$5x_1 + 4x_2 + 3x_3 \leq 11$$
$$3x_1 + 4x_2 + 2x_3 \leq 8$$

   holds. Derive the valid inequality

$$2x_1 + 2x_2 + x_3 \leq 4$$

   using the VP-MDFF $f_{I,FS,1}$ of Corollary 4.1, p. 102, with the parameter choice $\mathbf{v} := (3, 2)^\top$.

3. Let $\mathbf{w} := (1, \ldots, 1)^\top \in \mathbb{R}^m$. Given the inequality system

$$\sum_{j=1}^{n} \mathbf{a}^j x_j \leq \mathbf{w}; \; \mathbf{x} \geq \mathbf{0},$$

   where $\mathbf{a}^j \in [0, 1]^m$ for all j, suppose that a VP-MDFF $f : [0, 1]^m \to [0, 1]$ yields $f(\mathbf{a}^{j_0}) = 1$ for a certain $j_0 \in \{1, \ldots, n\}$. Why does this imply $f(\mathbf{a}^j) = 0$ for all those j, for which $\mathbf{a}^{j_0} + \mathbf{a}^j \leq \mathbf{w}$? What is the conclusion for the possible usage of the functions of Classes II, III and IV?

Appendix A
Hints and Solutions to Selected Exercises

Chapter 1

1.1. As stated in the text, the matrix $\mathbf{A}$ is just the vector $(1,3)$ and the vector $\mathbf{c} = (1,1)^\top$.

(a) The extreme points of X are $\mathbf{x}_1 = (0,0)^\top$, $\mathbf{x}_2 = (\frac{5}{2},0)^\top$, $\mathbf{x}_3 = (0,4)^\top$ with $X = \{\mathbf{x} \in \mathbb{R}^2 : \mathbf{x} = \lambda_1(0,0)^\top + \lambda_2(\frac{5}{2},0)^\top + \lambda_3(0,4)^\top, \lambda_1 + \lambda_2 + \lambda_3 = 1, \lambda_1, \lambda_2, \lambda_3 \geq 0\}$. Therefore, $[\mathbf{c}^\top\mathbf{x}_1, \mathbf{c}^\top\mathbf{x}_2, \mathbf{c}^\top\mathbf{x}_3] = [0, \frac{5}{2}, 4]$ and $[\mathbf{A}\mathbf{x}_1, \mathbf{A}\mathbf{x}_2, \mathbf{A}\mathbf{x}_3] = [0, \frac{5}{2}, 12]$. The DW-model is: $z_{DW} := \max\{0\lambda_1 + \frac{5}{2}\lambda_2 + 4\lambda_3 : 0\lambda_1 + \frac{5}{2}\lambda_2 + 12\lambda_3 \leq 6, \lambda_1 + \lambda_2 + \lambda_3 = 1, \lambda_1, \lambda_2, \lambda_3 \geq 0\}$. The optimal solution is $(\lambda_1, \lambda_2, \lambda_3)^* = (0, \frac{12}{19}, \frac{7}{19})$, which maps to the solution in the original space $\mathbf{x}^* = (\frac{30}{19}, \frac{28}{19})^\top$.

(b) $z_{IP}^* = 2$, $z_{DWI}^* = \frac{14}{5} = 2.8$, $z_{DW}^* = z_{LP}^* = \frac{58}{19} = 3.053$.

1.2. Check Fig. A.1 to identify the extreme points of the sets X^1 and X^2, respectively. The matrices and vectors in the reformulated model are as follows:

$$\mathbf{c}^{1\top}\mathbf{X}^1 = \mathbf{c}^{1\top}\begin{bmatrix}\mathbf{x}_1^1 & \mathbf{x}_2^1 & \mathbf{x}_3^1 & \mathbf{x}_4^1\end{bmatrix} = \begin{bmatrix}3 & 5\end{bmatrix} * \begin{bmatrix}0 & 2 & 2 & 0\\ 0 & 0 & 1 & 2\end{bmatrix} = \begin{bmatrix}0 & 6 & 11 & 10\end{bmatrix}$$

$$\mathbf{A}^1\mathbf{X}^1 = \mathbf{A}^1\begin{bmatrix}\mathbf{x}_1^1 & \mathbf{x}_2^1 & \mathbf{x}_3^1 & \mathbf{x}_4^1\end{bmatrix} = \begin{bmatrix}1 & 2\\ 3 & 2\end{bmatrix} * \begin{bmatrix}0 & 2 & 2 & 0\\ 0 & 0 & 1 & 2\end{bmatrix} = \begin{bmatrix}0 & 2 & 4 & 4\\ 0 & 6 & 8 & 4\end{bmatrix}$$

$$\mathbf{c}^{2\top}\mathbf{X}^2 = \mathbf{c}^{2\top}\begin{bmatrix}\mathbf{x}_1^2 & \mathbf{x}_2^2 & \mathbf{x}_3^2\end{bmatrix} = \begin{bmatrix}1 & 2\end{bmatrix} * \begin{bmatrix}0 & 1 & 0\\ 0 & 0 & 3\end{bmatrix} = \begin{bmatrix}0 & 1 & 6\end{bmatrix}$$

$$\mathbf{A}^2\mathbf{X}^2 = \mathbf{A}^2\begin{bmatrix}\mathbf{x}_1^2 & \mathbf{x}_2^2 & \mathbf{x}_3^2\end{bmatrix} = \begin{bmatrix}2 & 1\\ 1 & 1\end{bmatrix} * \begin{bmatrix}0 & 1 & 0\\ 0 & 0 & 3\end{bmatrix} = \begin{bmatrix}0 & 2 & 3\\ 0 & 1 & 3\end{bmatrix}.$$

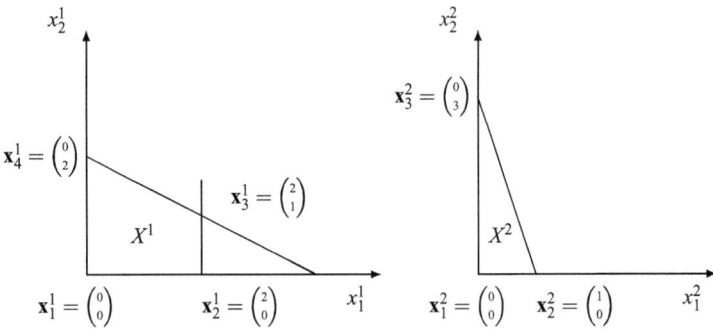

Fig. A.1 Sets X^1 and X^2

The original model with a block angular structure is reformulated into the following DW-model:

$$\begin{aligned}
\max z_{DW} := \quad & 6\lambda_2^1 + 11\lambda_3^1 + 10\lambda_4^1 \quad + 1\lambda_2^2 + 6\lambda_3^2 \\
\text{s. to} \quad & 2\lambda_2^1 + 4\lambda_3^1 + 4\lambda_4^1 \quad + 2\lambda_2^2 + 3\lambda_3^2 \leq 6 \\
& 6\lambda_2^1 + 8\lambda_3^1 + 4\lambda_4^1 \quad + 1\lambda_2^2 + 3\lambda_3^2 \leq 8 \\
& 1\lambda_1^1 + 1\lambda_2^1 + 1\lambda_3^1 + 1\lambda_4^1 \quad\quad\quad\quad\quad\quad\; = 1 \\
& \quad\quad\quad\quad\quad\quad\quad\quad\quad\; 1\lambda_1^2 + 1\lambda_2^2 + 1\lambda_3^2 = 1 \\
& \lambda_1^1, \lambda_2^1, \lambda_3^1, \lambda_4^1, \lambda_1^2, \lambda_2^2, \lambda_3^2 \geq 0.
\end{aligned}$$

The optimal solution is $(\lambda_1^1, \lambda_2^1, \lambda_3^1, \lambda_4^1, \lambda_1^2, \lambda_2^2, \lambda_3^2)^* = (0, 0, \frac{1}{2}, \frac{1}{2}, \frac{1}{3}, 0, \frac{2}{3})$, and $z_{DW}^* = \frac{29}{2}$. In terms of the original variables, the optimal solution is $(x_1^1, x_2^1, x_1^2, x_2^2)^* = (1, \frac{3}{2}, 0, 2)$, and $z^* = \frac{29}{2}$.

1.3. First, we address the decomposition with the constraints (1.18) in the subproblem.

(a) Let $X^i = \{\mathbf{x}_1^i, \ldots, \mathbf{x}_{k_i}^i\}$ be the set of all feasible assignments of jobs to machine i, $\forall i \in I$. A feasible assignment is a vector $\mathbf{x}_k^i = (x_{1k}^i, x_{2k}^i, \ldots, x_{|J|k}^i) \in \mathbb{B}^{|J|}$, whose elements indicate if the job is in the plan. The variables $y_k^i, k \in K_i, \forall i \in I$ in the reformulated model are defined as:

$$y_k^i = \begin{cases} 1, & \text{if feasible assignment } \mathbf{x}_k^i \text{ is used in machine } i \\ 0, & \text{otherwise.} \end{cases}$$

The reformulated model is as follows:

$$\max z := \sum_{i \in I, k \in K_i} (\sum_{j \in J} p_{ij} x_{jk}^i) y_k^i$$

Appendix A

$$\text{s. to} \sum_{i \in I, k \in K_i} \mathbf{x}^i_{jk} y^i_k = 1, \; j \in J$$

$$\sum_{k \in K_i} y^i_k \leq 1, \; i \in I$$

$$y^i_k \in \{0, 1\}, \; k \in K_i, i \in I$$

The block angular structure of the model leads to m knapsack subproblems, one for each machine. Concerning the other decomposition, with the constraints (1.19) in the subproblem, the reader may check that the reformulated model is the LP relaxation of the model (1.17)–(1.20).

(c) It is the model presented in Example 1.4.

(d) The lower bound is equal to 35. The reader may check that the optimal solution of the LP relaxation of the original is equal to 28.75.

1.4.

(b) The solution set of the subproblem is the set of all feasible circulation flows that are solutions of the set of constraints (1.22). A circulation flow is a flow along a directed cycle composed of arcs x_{ij} and the arc x_{L0}.
The only extreme point is the null solution, with $x_{ij} = 0, \forall (i,j) \in A$, and $x_{L0} = 0$.
Each extreme ray corresponds to the set of all feasible flows along a single path and the feedback arc x_{L0}. For instance, starting at the null solution, the points with $x_{04} = x_{48} = x_{80} = \theta$, for $\theta \geq 0$, define one of the extreme rays of the solution set.

(c) It is the model in Example 1.3.

1.5. The knapsack constraint for roll k, which defines a bounded set, is replaced by the convex hull of the integer extreme solutions to the knapsack polytope. These solutions are described by a vector $(x^p_{1k}, \ldots, x^p_{ik}, \ldots, x^p_{|I|k})^T, \forall k \in K, \forall p \in P$, where P is the set of feasible cutting patterns.

The reformulated model is as follows:

$$\min z := \sum_{k \in K} \sum_{p \in P} \lambda^p_k \quad \text{(A.1)}$$

$$\text{s. to} \sum_{k \in K} \sum_{p \in P} x^p_{ik} \lambda^p_k \geq d_i, \; \forall i \in I \quad \text{(A.2)}$$

$$\sum_{p \in P} \lambda^p_k \leq 1, \; \forall k \in K \quad \text{(A.3)}$$

$$\sum_{p \in P} x^p_{ik} \lambda^p_k \geq 0, \; \text{and integer}, \; \forall i \in I, \forall k \in K \quad \text{(A.4)}$$

$$\sum_{p \in P} \lambda_k^p \in \{0, 1\}, \ \forall k \in K \tag{A.5}$$

$$\lambda_k^p \geq 0, \forall p \in P, \ \forall k \in K, \tag{A.6}$$

where the decision variables λ_k^p denote the number of times pattern p is cut in roll k.

Constraints (A.2) guarantee that the demand for each item is satisfied by the patterns selected for the $|K|$ rolls. The convexity constraint for roll k (A.3) enforces that the solution is a convex combination of the extreme points of the knapsack polytope. The null solution is also an integer extreme point. As the corresponding column has the structure of a slack column, the convexity constraint is of the type $\leq$. This is consistent with the model, because some rolls may not be cut. When applying the decomposition, the integrality requirements on the y_k, $\forall k \in K$, in the original model, which are translated into constraints (A.5), cannot be dropped, as otherwise the integrality requirements (A.4) may not be sufficient to provide integer solutions to the cutting stock problem [see Valério de Carvalho (2002)].

The integrality requirements ensure that we pick a feasible cutting pattern for each roll k, and, as the rolls have equal length, the index k can be dropped and the solutions described by a vector $(x_1^p, \ldots, x_i^p, \ldots, x_{|I|}^p)^T$, $\forall p \in P$, leading to Gilmore and Gomory model [see Vance (1998) for further details].

Chapter 2

2.1. The function f_1 is a dual-feasible function, because $0 \leq f_1(x) \leq x$ for all $x \in [0, 1]$. f_2 is not a DFF, because $f_2(1/3) = 1/3$ and $f_2(2/3) = 1$, hence $f_2(1/3) + f_2(2/3) > 1$ in spite of $1/3 + 2/3 \leq 1$.

2.2. Choose any $x, y \in \mathbb{R}$. One gets $\lfloor f(x) \rfloor + \lfloor f(y) \rfloor \leq \lfloor f(x) + f(y) \rfloor \leq \lfloor f(x+y) \rfloor$, because $\lfloor \cdot \rfloor$ is non-decreasing.

2.3. The function f_3 is superadditive, while f_4 is not. The square of nonnegative numbers is superadditive, because $x, y \geq 0$ implies $(x+y)^2 - x^2 - y^2 = 2xy \geq 0$. The superadditivity of $x \mapsto x^2$ is preserved by the rounding-down. The function f_4 is not superadditive, because $f_4(0) > 0$.

2.4.

(a)

x	0	1	2	3	4	5	6	7	8	9	10	11	12	13	14	15	16	17	18	19
$g_1(x)$	0	0	0	0	1	2	3	4	5	5	5	5	6	7	8	9	10	10	10	10

(b) $f_5(x) = \begin{cases} 0, & \text{if } 0 \leq x < 4/19, \\ \lfloor 19x - 3 \rfloor/10, & \text{if } 4/19 \leq x \leq 8/19, \\ 1/2, & \text{if } 8/19 < x < 11/19, \\ \lfloor 19x - 6 \rfloor/10, & \text{if } 11/19 \leq x \leq 16/19, \\ 1, & \text{if } 16/19 < x \leq 1, \end{cases}$

Appendix A   137

(c) $f_6(x) = \begin{cases} 0, & \text{if } 0 \leq x < 3/19, \\ (19x-3)/10, & \text{if } 3/19 \leq x \leq 8/19, \\ 1/2, & \text{if } 8/19 < x < 11/19, \\ (19x-6)/10, & \text{if } 11/19 \leq x \leq 16/19, \\ 1, & \text{if } 16/19 < x \leq 1, \end{cases}$

(d) $f_7(x) = \begin{cases} 0, & \text{if } 0 \leq x < 4/19, \\ \lfloor 19x-3 \rfloor/10, & \text{if } 4/19 \leq x \leq 8/19, \\ 1/2, & \text{if } 8/19 < x < 11/19, \\ \lceil 19x-6 \rceil/10, & \text{if } 11/19 \leq x \leq 16/19, \\ 1, & \text{if } 16/19 < x \leq 1. \end{cases}$

2.5.

(a) False.
There are many counter examples. For instance, if $f \equiv f_{FS,1}(\cdot; 2)$ and $\ell_i < L/2$ for $i = 1, \ldots, m$ and $\mathbf{b} > \mathbf{o}$, then $\sum_{i=1}^{m} b_i \times f(\ell_i/L) = 0$, which is less than the continuous bound $\sum_{i=1}^{m} b_i \times \ell_i/L$, which is obtained by the identity function for g.

(b) True.
Setting $C := \frac{(k+1) \times kL}{kL+1}$ leads to $f_{BJ,1}(\frac{n}{L}; C) = f_{FS,1}(\frac{n}{L}; k)$ for $n = 0, 1, \ldots, L$.

Proof Both functions $f_{FS,1}$ and $f_{BJ,1}$ are maximal dual-feasible functions. Hence, the assertion needs to be verified for $0 < \frac{n}{L} < \frac{1}{2}$ only. Since $L, n \in \mathbb{N} \setminus \{0\}$, that allows the assumption $L \geq 3$.

Let $p \in \mathbb{N}$ with $p \leq k$ and $\frac{n}{L} \approx \frac{p}{k+1}$. Since $C = k + \frac{kL-k}{kL+1} \in (k, k+1)$, we obtain

$$f_{BJ,1}\left(\frac{n}{L}\right) = \left(\left\lfloor \frac{(k+1) \times kn}{kL+1} \right\rfloor + \max\left\{0, \frac{(kL+1) \times \text{frac}(\frac{(k+1) \times kn}{kL+1}) - kL + k}{1+k}\right\}\right)/k.$$

Three cases have now to be distinguished:

1. $n = \frac{Lp}{k+1} > 0$ yields

$$\frac{(k+1) \times kn}{kL+1} = \frac{kLp}{kL+1} = p - 1 + \frac{kL+1-p}{kL+1} \in (p-1, p),$$

and hence

$$f_{BJ,1}\left(\frac{n}{L}\right) = \left(p - 1 + \max\left\{0, \frac{kL+1-p-kL+k}{k+1}\right\}\right)/k$$

$$= \left(p - 1 + \frac{k+1-p}{k+1}\right)/k$$

$$= \frac{kp - k + p - 1 + k + 1 - p}{(k+1) \times k}$$

$$= \frac{p}{k+1}$$

$$= f_{FS,1}\left(\frac{p}{k+1}\right)$$

$$= f_{FS,1}\left(\frac{n}{L}\right).$$

2. $n < \frac{Lp}{k+1}$: Since k, L, n, p are integers, it follows that $n \leq \frac{Lp-1}{k+1}$. We get

$$C \times \frac{Lp-1}{(k+1) \times L} = \frac{k \times (Lp-1)}{kL+1} = p - 1 + \frac{kL + 1 - k - p}{kL + 1} \in (p-1, p),$$

because the nominator of the last fraction equals

$$k \times (L-1) + 1 - p \geq 2k + 1 - p > 0.$$

Therefore,

$$f_{BJ,1}\left(\frac{n}{L}\right) \leq f_{BJ,1}\left(\frac{Lp-1}{(k+1) \times L}\right)$$

$$= \left(p - 1 + \max\left\{0, \frac{kL + 1 - k - p - kL + k}{k+1}\right\}\right)/k$$

$$= \left(p - 1 + \max\left\{0, \frac{1-p}{k+1}\right\}\right)/k$$

$$= \frac{p-1}{k}$$

$$= f_{FS,1}\left(\frac{Lp-1}{(k+1) \times L}\right).$$

3. $n > \frac{Lp}{k+1}$: we get $n \geq \frac{Lp+1}{k+1}$ and

$$C \times \frac{Lp+1}{(k+1) \times L} = \frac{k \times (Lp+1)}{kL+1}$$

$$= p + \frac{k-p}{kL+1} \in [p, p+1),$$

Appendix A

and hence

$$f_{BJ,1}\left(\frac{n}{L}\right) \geq f_{BJ,1}\left(\frac{Lp+1}{(k+1) \times L}\right)$$

$$= \left(p + \max\left\{0, \frac{k-p-kL+k}{k+1}\right\}\right)/k$$

$$= \left(p + \max\left\{0, \frac{k \times (2-L) - p}{k+1}\right\}\right)/k$$

$$= \frac{p}{k}$$

$$= f_{FS,1}\left(\frac{Lp+1}{(k+1) \times L}\right).$$

We get for $\frac{Lp}{k+1} < n < \frac{L \times (p+1)}{k+1}$ from the combination of the second and third case $\frac{p}{k} \leq f_{BJ,1}\left(\frac{n}{L}\right) \leq \frac{p+1-1}{k}$, and hence $f_{BJ,1}\left(\frac{n}{L}\right) = \frac{p}{k}$ and analogously $\frac{p}{k} = f_{FS,1}\left(\frac{n}{L}\right)$ because of the monotonicity. □

(c) True.

For any $k \in \mathbb{N}$ with $k \geq 2$ and any $L \in \mathbb{N} \setminus \{0\}$, the use of $C := k - \frac{1}{L}$ leads to $f_{VB,2}\left(\frac{n}{L}; k\right) = f_{CCM,1}\left(\frac{n}{L}; C\right)$ for all $n \in \mathbb{N}$ with $n \leq L$.

Proof Since $f_{VB,2}$ and $f_{CCM,1}$ are maximal dual-feasible functions, it is enough to verify the proposition for $0 < n < L/2$. Then, it follows that

$$f_{VB,2}\left(\frac{n}{L}\right) = \left\lceil \frac{kn}{L} - 1 \right\rceil / (k-1)$$

and

$$f_{CCM,1}\left(\frac{n}{L}\right) = \left\lfloor \frac{Cn}{L} \right\rfloor / (k-1).$$

For any $x \in \mathbb{R} \setminus \mathbb{Z}$, it holds that $\lceil x - 1 \rceil = \lfloor x \rfloor$. One has

$$\frac{Cn}{L} = \frac{kn}{L} - \frac{n}{L^2} \in \left(\frac{kn-1}{L}, \frac{kn}{L}\right),$$

because $0 < n < \frac{L}{2}$.

If $L | kn$, then

$$f_{VB,2}\left(\frac{n}{L}\right) = \left(\frac{kn}{L} - 1\right)/(k-1)$$

and

$$\left\lfloor \frac{Cn}{L} \right\rfloor = \frac{kn}{L} - 1,$$

i.e.,

$$f_{CCM,1}\left(\frac{n}{L}\right) = \left(\frac{kn}{L} - 1\right)/(k-1) = f_{VB,2}\left(\frac{n}{L}\right).$$

If $L \nmid kn$, then $f_{VB,2}\left(\frac{n}{L}\right) = \lfloor \frac{kn}{L} \rfloor/(k-1) = \lfloor \frac{kn-1}{L} \rfloor/(k-1) = f_{CCM,1}\left(\frac{n}{L}\right)$. $\square$

(d) False.

Choose any $p \in \mathbb{N}$ with $p \geq 3$. Let $k := 3$, $L := 3p$, $m \geq 2$, $\ell_1 := p$ and $\ell_2 := p+1$. Suppose there is a $C \in \mathbb{R}$ with $C \geq 1$ and

$$f_{VB,2}\left(\frac{p+1}{3p}; 3\right) = f_{BJ,1}\left(\frac{p+1}{3p}; C\right)$$

and

$$f_{VB,2}\left(\frac{p}{3p}; 3\right) = f_{BJ,1}\left(\frac{1}{3}; C\right).$$

The latter yields

$$f_{VB,2}\left(\frac{1}{3}; 3\right) = 0 = \left\lfloor \frac{C}{3} \right\rfloor,$$

i.e.

$$1 \leq C < 3, \quad \text{and} \quad \frac{C}{3} \leq \text{frac}(C).$$

The case $2 \leq C < 3$ leads to $\text{frac}(C) = C - 2 \geq \frac{C}{3}$, and hence $\frac{2}{3}C \geq 2$, implying the contradiction $C \geq 3$. Therefore, $1 \leq C < 2$ and $\text{frac}(C) = C - 1 \geq \frac{C}{3}$ such that $\frac{3}{2} \leq C < 2$. One has

$$f_{BJ,1}\left(\frac{p+1}{3p}; C\right) = \left\lfloor \frac{p+1}{3p} \times C \right\rfloor + \max\left\{0, \frac{\text{frac}(\frac{p+1}{3p} \times C) - C + 1}{2 - C}\right\}$$

$$= 0 + \max\left\{0, \frac{\frac{p+1}{3p} \times C - C + 1}{2 - C}\right\}.$$

Appendix A

Since $p \geq 3$ and $\frac{3}{2} \leq C < 2$, one gets $(p-2) \times (3-2C) \leq 0$, and hence $p \times (3-2C) \leq 6-4C$. Therefore, $C \times (1-2p) + 3p \leq 6-3C$ and

$$\frac{\frac{p+1}{3p} \times C - C + 1}{2 - C} = \frac{C \times (p+1) - 3Cp + 3p}{3p \times (2-C)}$$

$$= \frac{C \times (1-2p) + 3p}{p \times (6-3C)}$$

$$\leq \frac{1}{p}$$

in contradiction to $f_{VB,2}\left(\frac{p+1}{3p}; 3\right) = \frac{1}{2}$.

2.6. The proof is given next.

Proof One has obviously $f_{BJ,1}(0) = 0$ and $f_{BJ,1}(x) \geq 0$ for all $x \in [0,1]$. To show the symmetry, choose any $x \in [0,1]$. We get either $\mathsf{frac}(Cx) \leq \mathsf{frac}(C)$ and hence

$$\mathsf{frac}(C - Cx) = \mathsf{frac}(C) - \mathsf{frac}(Cx) \leq \mathsf{frac}(C),$$

$$f_{BJ,1}(x) = \lfloor Cx \rfloor / \lfloor C \rfloor,$$

$$f_{BJ,1}(1-x) = \lfloor C - Cx \rfloor / \lfloor C \rfloor$$

$$= (\lfloor C \rfloor - \lfloor Cx \rfloor) / \lfloor C \rfloor,$$

or one has $\mathsf{frac}(Cx) > \mathsf{frac}(C)$, and consequently $\mathsf{frac}(C - Cx) = \mathsf{frac}(C) + 1 - \mathsf{frac}(Cx) > \mathsf{frac}(C)$ and

$$f_{BJ,1}(x) + f_{BJ,1}(1-x) =$$

$$= \left(\lfloor Cx \rfloor + \lfloor C - Cx \rfloor + \frac{\mathsf{frac}(Cx) - \mathsf{frac}(C)}{1 - \mathsf{frac}(C)} + \frac{\mathsf{frac}(C - Cx) - \mathsf{frac}(C)}{1 - \mathsf{frac}(C)}\right) / \lfloor C \rfloor$$

$$= \left(\lfloor Cx \rfloor + \lfloor C \rfloor - \lfloor Cx \rfloor - 1 + \right.$$

$$\left. \frac{\mathsf{frac}(Cx) - \mathsf{frac}(C) + \mathsf{frac}(C) + 1 - \mathsf{frac}(Cx) - \mathsf{frac}(C)}{1 - \mathsf{frac}(C)}\right) / \lfloor C \rfloor$$

$$= 1 + \left(\frac{1 - \mathsf{frac}(C)}{1 - \mathsf{frac}(C)} - 1\right) / \lfloor C \rfloor$$

$$= 1.$$

To show the superadditivity, take any $x, y \in [0,1]$ with $x + y \leq 1$ and, without loss of generality, take also $\mathsf{frac}(x) \leq \mathsf{frac}(y)$. Let

$$d := (f_{BJ,1}(x+y) - f_{BJ,1}(x) - f_{BJ,1}(y)) \times \lfloor C \rfloor.$$

We have to show $d \geq 0$. Two cases are distinguished.

- If $\text{frac}(Cx) + \text{frac}(Cy) < 1$, then $\lfloor Cx \rfloor + \lfloor Cy \rfloor = \lfloor Cx + Cy \rfloor$ and $\text{frac}(Cx+Cy) = \text{frac}(Cx) + \text{frac}(Cy)$, such that

$$d = \max\left\{0, \frac{\text{frac}(Cx) + \text{frac}(Cy) - \text{frac}(C)}{1 - \text{frac}(C)}\right\} - \max\left\{0, \frac{\text{frac}(Cx) - \text{frac}(C)}{1 - \text{frac}(C)}\right\} - \max\left\{0, \frac{\text{frac}(Cy) - \text{frac}(C)}{1 - \text{frac}(C)}\right\}.$$

If $\text{frac}(Cx) \leq \text{frac}(C)$, then $d \geq 0$ becomes obvious. Otherwise

$$d = \frac{\text{frac}(Cx) + \text{frac}(Cy) - \text{frac}(C)}{1 - \text{frac}(C)} - \frac{\text{frac}(Cx) - \text{frac}(C)}{1 - \text{frac}(C)} - \frac{\text{frac}(Cy) - \text{frac}(C)}{1 - \text{frac}(C)}$$

$$= \frac{\text{frac}(C)}{1 - \text{frac}(C)}$$

$$\geq 0.$$

- If $\text{frac}(Cx) + \text{frac}(Cy) \geq 1$, then $\lfloor Cx \rfloor + \lfloor Cy \rfloor = \lfloor Cx + Cy \rfloor - 1$ and $\text{frac}(Cx + Cy) = \text{frac}(Cx) + \text{frac}(Cy) - 1$. Hence, we have

$$d = \lfloor C \rfloor + \max\left\{0, \frac{\text{frac}(Cx) + \text{frac}(Cy) - 1 - \text{frac}(C)}{1 - \text{frac}(C)}\right\}$$

$$- \max\left\{0, \frac{\text{frac}(Cx) - \text{frac}(C)}{1 - \text{frac}(C)}\right\} - \max\left\{0, \frac{\text{frac}(Cy) - \text{frac}(C)}{1 - \text{frac}(C)}\right\}.$$

If $\text{frac}(Cx) \leq \text{frac}(C)$, then $d > \lfloor C \rfloor - 1 \geq 0$, because $\frac{\text{frac}(Cy) - \text{frac}(C)}{1 - \text{frac}(C)} < \frac{1 - \text{frac}(C)}{1 - \text{frac}(C)} = 1$. If $\text{frac}(Cx) > \text{frac}(C)$, then

$$d \geq \lfloor C \rfloor + \frac{\text{frac}(Cx) + \text{frac}(Cy) - 1 - \text{frac}(C)}{1 - \text{frac}(C)}$$

$$- \frac{\text{frac}(Cx) - \text{frac}(C)}{1 - \text{frac}(C)} - \frac{\text{frac}(Cy) - \text{frac}(C)}{1 - \text{frac}(C)}$$

$$= \lfloor C \rfloor + \frac{\text{frac}(C) - 1}{1 - \text{frac}(C)}$$

$$= \lfloor C \rfloor - 1$$

$$\geq 0.$$

□

Appendix A

2.7. The function f is a maximal dual-feasible function, but not extreme, because it is symmetric and strict convex on $[0, 1/2]$ and also repeatedly differentiable in $(0, 1/2)$.

2.8. The proof is given next.

Proof Let $g, h : [0, 1] \to [0, 1]$ be maximal dual-feasible functions with $2 \times f_{FS,1}(x) = g(x) + h(x)$ for all $x \in [0, 1]$. One has to show that $f_{FS,1}(x) = g(x)$ for all $x \in (0, 1/2)$ with $0 < f_{FS,1}(x) < 1/2$. Therefore assume $k > 1$. Since g, h are dual-feasible functions, it follows that

$$(k+1) \times g\left(\frac{1}{k+1}\right) \leq 1$$

and

$$h\left(\frac{1}{k+1}\right) \leq \frac{1}{k+1}.$$

Since $f_{FS,1}\left(\frac{1}{k+1}\right) = \frac{1}{k+1}$, one gets $g\left(\frac{1}{k+1}\right) = \frac{1}{k+1}$, implying $g\left(\frac{p}{k+1}\right) \geq \frac{p}{k+1}$ for any $p \in \{1, 2, \ldots, k\}$ due to the monotonicity of g. Analogously $h\left(\frac{p}{k+1}\right) \geq \frac{p}{k+1}$, and hence $g(\frac{p}{k+1}) = f(\frac{p}{k+1})$. It remains to show $g(x) = f(x)$ if $(k+1) \times x \notin \mathbb{N}$. There is a $p \in \mathbb{N}$ with

$$0 < p < x \times (k+1) < p+1$$

and $p < k/2$, such that $f_{FS,1}(x) = p/k$. Let $x_0 := p/k$. Because of $0 < p < k/2 < k$, it holds that

$$p < p \times (k+1)/k = x_0 \times (k+1) < p+1,$$

i.e. $f_{FS,1}(x_0) = x_0$. Similarly to the other case $g\left(\frac{1}{k}\right) = \frac{1}{k} = f_{FS,1}\left(\frac{1}{k}\right)$ and $g(x_0) = x_0$ follows. Since x_0 is in the inner of an open interval on which $f_{FS,1}$ is constant, g must be constant in the same interval. Otherwise the monotonicity of g and h would yield a contradiction. This implies

$$g(x) = g(x_0) = x_0 = f_{FS,1}(x_0) = f_{FS,1}(x)$$

for all $x \in (\frac{p}{k+1}, \frac{p+1}{k+1})$, and hence $g \equiv f_{FS,1}$. □

2.9.

(a) True.

If f is not extreme, then there are different maximal dual-feasible functions $f_1, f_2 : [0, 1] \to [0, 1]$ with $2f \equiv f_1 + f_2$. There is an $x_0 \in [0, 1]$ with $f_1(x_0) \neq f_2(x_0)$. Since g is surjective, there is an $x_1 \in [0, 1]$ with $g(x_1) = x_0$ that yields $f_1(g(x_1)) = f_1(x_0) \neq f_2(x_0) = f_2(g(x_1))$. Hence, $f_1(g(\cdot))$ and

$f_2(g(\cdot))$ are different functions. Moreover, for all $x \in [0,1]$, it holds that
$2f(g(x)) - f_1(g(x)) - f_2(g(x)) = 0$.

(b) True.
If f is continuous, then the image of the interval $[0, 1]$, which is a connected set, must be connected, and hence an interval. Since $f(0) = 0$ and $f(1) = 1$, this image is the interval $[0, 1]$, such that f is surjective. The opposite direction consists in showing the continuity of f, provided that f is surjective. The function f is continuous at the point zero, because for all $x \in (0, 1]$, it holds that

$$0 \leq f(x) \leq 1/\lfloor 1/x \rfloor.$$

Hence, $\lim_{x \downarrow 0} f(x) = 0$. The symmetry of f implies also the continuity at 1. Let $\bar{x} \in (0, 1)$ be an arbitrary constant. Let (x_n) and (y_n) be any sequences with

$$0 \leq x_1 \leq x_2 \leq \cdots \leq \bar{x} \leq \cdots \leq y_2 \leq y_1 \leq 1$$

and

$$\lim_{n \to \infty} x_n = \lim_{n \to \infty} y_n = \bar{x}.$$

The monotonicity of f implies

$$0 \leq f(x_1) \leq f(x_2) \leq \cdots \leq f(\bar{x}) \leq \cdots \leq f(y_2) \leq f(y_1) \leq 1.$$

Every monotone and bounded sequence converges, hence the left and right limits of f at $\bar{x}$ exist, and it holds that

$$\lim_{x \uparrow \bar{x}} f(x) \leq f(\bar{x}) \leq \lim_{x \downarrow \bar{x}} f(x).$$

Since f is surjective, it cannot happen that $\lim_{x \uparrow \bar{x}} f(x) < f(\bar{x})$ or $\lim_{x \downarrow \bar{x}} f(x) > f(\bar{x})$.

(c) False.
The function $f_{FS,1}(\cdot; 1)$ defined in (2.10) (p. 35), is a counter example. This function is an extreme maximal dual-feasible function and convex on $[0, 1/2]$, but not continuous.

2.10.

(a) The function $f_{MT,0}$ is obviously symmetric, monotone and nonnegative. According to Theorem 2.2, only the superadditivity remains to be proved. Choose any $x, y \in (0, 1/2)$ with $x \leq y$. If $x < \lambda$, then $f_{MT,0}(x) = 0$, such that the

Appendix A

superadditivity follows from the monotonicity. If $x \geq \lambda$, then

$$f_{MT,0}(x) + f_{MT,0}(y) = x + y \leq f_{MT,0}(x + y).$$

(b) Let $g, h : [0, 1] \to [0, 1]$ be any maximal dual-feasible functions with $2f_{MT,0} \equiv g + h$. We show $g(x) = h(x)$ for all x. If $x < \lambda$, then $f_{MT,0}(x) = 0$ implies immediately $g(x) = h(x) = 0$ due to the nonnegativity of g and h. It holds for all $x, y \in [\lambda, \frac{1-\lambda}{2}]$ that

$$f_{MT,0}(x + y) = x + y = f_{MT,0}(x) + f_{MT,0}(y).$$

Since $f_{MT,0}(1/4) = 1/4$ and $f_{MT,0}(1/3) = 1/3$, it follows that $g(1/4) = h(1/4) = 1/4$ and $g(1/3) = h(1/3) = 1/3$, because larger function values would violate the definition of a dual-feasible function. Now Lemma 2.4 can be applied with $a := 1/4$ and $b := 1/3$, yielding $g(x) = h(x) = x$ for all $x \in [\frac{1}{4}, \frac{1}{3}]$. The superadditivity of g and h implies $g(2x) \geq 2g(x) = 2x$ and $h(2x) \geq 2x$ for all these x, and hence $g(2x) = h(2x) = 2x$. The symmetry yields $g(x) = h(x) = x$ for all $x \in [\frac{1}{3}, \frac{1}{2}] \cup [\frac{2}{3}, \frac{3}{4}]$ too, and hence for all $x \in [\frac{1}{4}, \frac{3}{4}]$. Moreover, for $x \in (\frac{3}{4}, 1 - \lambda]$, one obtains $g(x/2) = h(x/2) = x/2$ and again due to the superadditivity $g(x) = h(x) = x$. Finally the symmetry yields $g \equiv h$. □

2.11. The proof is given next.

Proof Assume without loss of generality $\alpha \leq \beta$. If $\alpha = \beta$, then, according to Definition 2.6, the different maximal dual-feasible functions f, g immediately yield that h is not extreme. Otherwise define the function $h_1 : [0, 1] \to [0, 1]$ by

$$h_1(x) := 2h(x) - g(x) = \frac{2\alpha}{\alpha + \beta} f(x) + \frac{\beta - \alpha}{\alpha + \beta} g(x).$$

That yields $2h \equiv g + h_1$, i.e. if h_1 is proved to be a maximal dual-feasible function different from g, then the proof is complete. Since $0 < \alpha < \beta$, both factors $\frac{2\alpha}{\alpha+\beta}$ and $\frac{\beta-\alpha}{\alpha+\beta}$ are positive, such that h_1 is a maximal dual-feasible function due to Proposition 2.2. Since $f \neq g$, there is an $x_0 \in [0, 1]$ with $f(x_0) \neq g(x_0)$. Suppose $h_1(x_0) = g(x_0)$. That assumption would imply

$$\frac{2\alpha}{\alpha + \beta} f(x_0) = \frac{2\alpha}{\alpha + \beta} g(x_0)$$

in contradiction to $f(x_0) \neq g(x_0)$ and $\alpha > 0$. □

2.12. Consider the following functions

$$f(x) := \begin{cases} 1/2, & \text{if } x = 2/5, \\ 1, & \text{if } x = 2/3, \\ 0, & \text{otherwise,} \end{cases}$$

and

$$g(x) := \begin{cases} 2/5, & \text{if } x = 1/2, \\ 2/3, & \text{if } 1/2 < x \leq 1, \\ 0, & \text{otherwise.} \end{cases}$$

Both f and g are dual-feasible functions, but neither symmetric nor superadditive. Their composition yields

$$f(g(x)) = f_{FS,1}(x;1) = \begin{cases} 0, & \text{if } 0 \leq x < 1/2, \\ 1/2, & \text{if } x = 1/2, \\ 1, & \text{if } 1/2 < x \leq 1. \end{cases}$$

for any $x \in [0,1]$.

2.13. The function $f_{LL,1}$ is symmetric if and only if $k = 2$.

Proof If $k = 2$, then choose any $x \in [0,1]$. If $\text{frac}(Cx) \leq \text{frac}(C)$, then $\text{frac}(C - Cx) \leq \text{frac}(C)$, and hence

$$f_{LL,1}(x) + f_{LL,1}(1-x) = \frac{\lfloor Cx \rfloor + \lfloor C - Cx \rfloor}{\lfloor C \rfloor}$$

$$= \frac{\lfloor Cx \rfloor + \lfloor C \rfloor - \lfloor Cx \rfloor}{\lfloor C \rfloor}$$

$$= 1.$$

If $\text{frac}(Cx) > \text{frac}(C)$, then $\text{frac}(C - Cx) > \text{frac}(C)$ and

$$f_{LL,1}(x) + f_{LL,1}(1-x) = \left(\lfloor Cx \rfloor + \lfloor C - Cx \rfloor + \left\lceil \frac{\text{frac}(Cx) - \text{frac}(C)}{1 - \text{frac}(C)} \right\rceil / 2 \right.$$

$$\left. + \left\lceil \frac{\text{frac}(C - Cx) - \text{frac}(C)}{1 - \text{frac}(C)} \right\rceil / 2 \right) / \lfloor C \rfloor$$

$$= (\lfloor Cx \rfloor + \lfloor C \rfloor - 1 - \lfloor Cx \rfloor + 1/2 + 1/2) / \lfloor C \rfloor$$

$$= 1.$$

Appendix A   147

If $k > 2$, then let $x := \left(\text{frac}(C) + \frac{1-\text{frac}(C)}{k-1}\right)/C$. Then $x > 0$ is obvious. It holds also that $\frac{1-\text{frac}(C)}{k-1} < 1$, and hence $x < (\text{frac}(C) + 1)/C \leq 1$. We show

$$f_{LL,1}(x) + f_{LL,1}(1-x) < 1.$$

First, one gets

$$Cx = \text{frac}(C) + \frac{1-\text{frac}(C)}{k-1} \leq \text{frac}(C) + \frac{1-\text{frac}(C)}{2}$$
$$= \frac{1+\text{frac}(C)}{2} < 1$$

and $\text{frac}(Cx) > \text{frac}(C)$. That implies

$$0 < \text{frac}(C - Cx) - \text{frac}(C) = 1 - \text{frac}(Cx) = 1 - \text{frac}(C) - \frac{1-\text{frac}(C)}{k-1}$$
$$= \frac{1-\text{frac}(C)}{k-1} \times (k-2)$$

and

$$f_{LL,1}(x) + f_{LL,1}(1-x) = \left(\lfloor Cx \rfloor + \lfloor C - Cx \rfloor + \left\lceil \frac{\text{frac}(Cx) - \text{frac}(C)}{1-\text{frac}(C)} \times (k-1) \right\rceil / k\right.$$
$$\left. + \left\lceil \frac{\text{frac}(C-Cx) - \text{frac}(C)}{1-\text{frac}(C)} \times (k-1) \right\rceil / k \right)/\lfloor C \rfloor$$
$$= \left(\lfloor Cx \rfloor + \lfloor C \rfloor - 1 - \lfloor Cx \rfloor + 1/k + \frac{k-2}{k}\right)/\lfloor C \rfloor$$
$$= 1 - \frac{1}{k \times \lfloor C \rfloor} < 1.$$

□

2.14. The proof is given next.

Proof Let $x := 1/k$. Then, $x \in [0, 1]$ and $f_{VB,1}(x) = 0$, and hence $f_{VB,1}(1-x) = \frac{k-2}{k-1} < 1$.

□

2.15.

(a) False.
    Let for example $C := 1.9$, $x := 0.3$ and g be the maximal dual-feasible function (2.10) with $k = 2$. Then, $f(x) = g(x) = 1, f(2x) = 1 < 2 * f(x)$.

(b) False.

Let *e.g.* $C := 1$, $x := -8$, $y := 2$ and g be the identity function. One gets $f(x) = 0, f(y) = 1, f(x+y) = f(-6) = 0 < f(x) + f(y)$.

(c) True.

If $x_1, \ldots, x_n \in \mathbb{R}_+$ with $\sum_{i=1}^{n} x_i \leq \frac{1}{C}$ are given, then using each summand C times, i.e. setting $x_{np+i} := x_i$ for $i = 1, \ldots, n$ and $p = 1, \ldots, C-1$, yields

$$\sum_{i=1}^{Cn} x_i = C \times \sum_{i=1}^{n} x_i \leq 1.$$

Hence, according to Definition 2.1, we have

$$\sum_{i=1}^{Cn} f(x_i) = C \times \sum_{i=1}^{n} x_i \leq 1.$$

Chapter 3

3.1. The proof is given next.

Proof If f is a general dual-feasible function and $x \in (0, 1]$, then let $n := \lfloor 1/x \rfloor \in \mathbb{N} \setminus \{0\}$. Setting

$$x_1 := x_2 := \cdots := x_n := x$$

yields

$$\sum_{i=1}^{n} x_i = nx \leq 1$$

and

$$\sum_{i=1}^{n} f(x_i) = n \times f(x) \leq 1$$

due to Definition 3.1. Hence, we have $f(x) \leq 1/n$.

Given $n \in \mathbb{N}$ and $x_i \in \mathbb{R}$ ($i = 1, \ldots, n$) with $\sum_{i=1}^{n} x_i \leq 0$, it cannot happen that

$$\sum_{i=1}^{n} f(x_i) = \varepsilon > 0,$$

because using each x_i in the quantity $p := \lfloor 1/\varepsilon \rfloor + 1$, i.e. setting

$$x_{n+1} := x_1, \ldots, x_{np} := x_n,$$

one gets

$$\sum_{i=1}^{np} x_i = p \times \sum_{i=1}^{n} x_i \leq 0,$$

but

$$\sum_{i=1}^{np} f(x_i) = p \times \varepsilon > 1$$

in contradiction to Definition 3.1, and the fact that f is a dual-feasible function. □

3.2. Suppose, there is an $x_1 < 0$ with $f(x_1) \geq 0$. Let $n := \lfloor 1/f(x_0) \rfloor + 1$. Then, $n \in \mathbb{N} \setminus \{0\}$ and $n \times f(x_0) > 1$. Taking n times the summand x_0 and m times the summand x_1, where $m \in \mathbb{N}$ is chosen appropriately, one gets $n \times x_0 + m \times x_1 \leq 1$, but $n \times f(x_0) + m \times f(x_1) > 1$ in contradiction to Definition 3.1.

3.3. The sequence $x_2, \ldots, x_n$ can contain x_1 repeatedly, even if $x_1 > 1/2$, because negative summands may occur too. Suppose that $x_1 = x_2 = \cdots = x_k$, with $k \in \mathbb{N}$ and $0 < k \leq n$. Instead of

$$\sum_{i=2}^{n} f(x_i) > f(1 - x_1),$$

one gets

$$1 < \sum_{i=1}^{n} h(x_i)$$

$$= k \times h(x_1) + \sum_{i=k+1}^{n} h(x_i)$$

$$= k - k \times f(1 - x_1) + \sum_{i=k+1}^{n} f(x_i),$$

but this does not imply a contradiction.

3.4.

(a) 353/122.

(b) Consider for instance the following four patterns that are cut once each:

$$(1,0,1,0,0,0)^\top,$$
$$(1,0,0,1,0,2)^\top,$$
$$(0,1,0,0,2,1)^\top,$$
$$(0,0,0,0,0,1)^\top.$$

The corresponding waste lengths are equal to 10, 6, 9 and 110.

(c) The pattern

$$(1,1,0,0,0,0)^\top$$

needs the length 123. Use this and for example the following two patterns once:

$$(1,0,1,0,0,0)^\top \quad (\text{waste}\,10), \text{ and}$$
$$(0,0,0,1,2,4)^\top \quad (\text{waste}\,4).$$

(d)

C	$\frac{61}{55}$	$\frac{61}{51}$	$\frac{61}{46}$	$\frac{61}{36}$	$\frac{61}{31}$	2	$\frac{61}{30}$	$\frac{61}{25}$	$\frac{61}{15}$	$\frac{61}{10}$	$\frac{61}{6}$
Without extra length	$\frac{221}{98}$	$\frac{165}{82}$	$\frac{115}{62}$	$\frac{35}{22}$	$\frac{5}{2}$	$\frac{353}{122}$	$\frac{163}{58}$	$\frac{15}{7}$	$\frac{131}{56}$	$\frac{77}{27}$	$\frac{143}{50}$
With extra length	$\frac{110}{49}$	2	$\frac{57}{31}$	$\frac{17}{11}$	2	$\frac{176}{61}$	$\frac{325}{116}$	$\frac{17}{8}$	$\frac{261}{112}$	$\frac{307}{108}$	$\frac{57}{20}$

Without extra space, the largest lower bound was $\frac{353}{122} \approx 2.89$. The extra space counts like an item of length -1. Its contribution decreases the obtained lower bound only slightly to $\frac{176}{61}$.

Remark Here, the optimal objective function value of the continuous relaxation is a bit larger than the material bound, namely $2\frac{32}{35} \approx 2.91$ (without extra space).

3.5.

(a) False.
Non-maximal general dual-feasible functions need not to have this structure. An example is

$$x \mapsto -e^{|x|}.$$

(b) True.
Let $f : \mathbb{R} \to \mathbb{R}$ be any general dual-feasible function and $g : \mathbb{R} \to \mathbb{R}$ a maximal general dual-feasible function dominating f. (If f is already maximal,

then $g = f$.) Proposition 3.4 (p. 62) implies

$$f(x) \leq g(x) \leq tx$$

for all $x \in \mathbb{R}$, where $t := \sup_{x>0} \frac{f(x)}{x}$ remains finite.

3.6. The proof is given next.

Proof We show first that $\mathscr{L} \subseteq \mathscr{F}$ similarly to Proposition 3.2, p. 54. Let $f \in \mathscr{L}$ and therefore the constant c be given. One gets for any finite index set I of real numbers

$$\sum_{i \in I} x_i \leq 0 \implies \sum_{i \in I} f(x_i) = c \times \sum_{i \in I} x_i \leq 0,$$

such that f fulfills the first condition. Assume that f is dominated by a real function g, i.e. $f(x) \leq g(x)$ for all $x \in \mathbb{R}$ and there is an $y \in \mathbb{R}$ with $g(y) > f(y)$. Then,

$$g(-y) \geq f(-y)$$

and

$$g(y) + g(-y) > f(y) + f(-y) = 0,$$

such that g does not fulfill the implication (3.2).

It remains to show that $\mathscr{F} \subseteq \mathscr{L}$. Let $f \in \mathscr{F}$ be given. One gets $f(0) = 0$,

$$f(x) \geq 0 \geq f(-x)$$

for all $x \in \mathbb{R}_+$, and also, analogously to part (b) of Theorem 3.1, that f is superadditive. Moreover,

$$c := \lim_{x \to \infty} \frac{f(x)}{x}$$

must be finite, because

$$f(x) + \lceil x \rceil \times f(-1) \leq 0$$

for all $x > 0$. Of course, we have $c \geq 0$. It follows that $f(x) \leq cx$ for all $x > 0$, otherwise the superadditivity of f would contradict the definition of c. It can also not happen that $f(x) > cx$ for a certain $x < 0$. A detailed proof could be done like in Proposition 3.4. Therefore, the linear function $x \mapsto cx$ dominates f, but f is maximal in the sense of the condition (2.), hence $f(x) = cx$ for all $x \in \mathbb{R}$. □

3.7. The functions $f_0, \ldots, f_3$ fulfill obviously the conditions (1.) and (3.) of Theorem 3.1. Moreover, f_0 is a superadditive general dual-feasible function, but not symmetric, and f_0 is not a maximal general dual-feasible functions. The reasons are the following:

$$f_0(1) = 1,$$

such that the superadditivity of f_0 will imply that f_0 is a general dual-feasible function. One gets for $x > 0$ that f_0 is differentiable and has the derivative

$$f_0'(x) = 1 + \tanh 1 - (1 - \tanh^2 x) = \tanh 1 + \tanh^2 x,$$

which rises strictly monotonely. Therefore, f_0 is strict convex for $x > 0$ and hence superadditive for positive arguments. The superadditivity holds without restriction, because if $x < 0 \leq y$, then

$$f_0(x+y) - f_0(x) - f_0(y) = \tanh y - \max\{0, \tanh(x+y)\} \geq 0,$$

and if $x, y < 0$, then $f_0(x+y) = f_0(x) + f_0(y)$. The strict convexity implies also $f_0(x) < x$ for $0 < x < 1$, such that f_0 cannot be symmetric and is dominated by $g : \mathbb{R} \to \mathbb{R}$ with

$$g(x) := \begin{cases} (1 + \tanh 1) \times x, & \text{if } x \leq 0, \\ x, & \text{if } 0 \leq x \leq 1, \\ (1 + \tanh 1) \times x - \tanh 1, & \text{if } x \geq 1, \end{cases}$$

which is a maximal general dual-feasible function according to Proposition 3.11 (with $p := 1$ and $t = 1 + \tanh 1$).

Let $\varepsilon > 0$ be sufficiently small. The functions f_1 and f_2 are not monotone, hence not superadditive, because one gets

$$f_1(2k) = f_2(2k) = 2k,$$

but

$$f_1(2k - \varepsilon) = f_2(2k - \varepsilon) = ((k+1) \times 2k - 1)/k = 2k + 2 - 1/k > 2k.$$

The function f_2 is not even a general dual-feasible function, because

$$f_2\left(\frac{-1}{k+1}\right) = \frac{-1}{k+1}$$

and

$$f_2(1 + \varepsilon) = 1 + \frac{1}{k},$$

Appendix A

yielding

$$f_2(\frac{-1}{k+1}) + f_2(1+\varepsilon) > 1$$

in spite of

$$1 + \varepsilon - \frac{1}{k+1} < 1.$$

The functions f_2 and f_3 are symmetric, but not f_1. One has for example $f_1(2) = 2$, but $f_1(-1) = -1 - 1/k$, and hence

$$f_1(-1) + f_1(2) = 1 - 1/k < 1.$$

Regarding f_2, if $(k+1) \times x \notin \mathbb{Z}$, then

$$\begin{aligned} f_2(x) + f_2(1-x) &= \lfloor (k+1) \times x \rfloor / k + \lfloor (k+1) \times (1-x) \rfloor / k \\ &= \lfloor (k+1) \times (x+1-x) - 1 \rfloor / k \\ &= 1. \end{aligned}$$

We show that f_1 is a general dual-feasible function. Let any finite index set I of real numbers x_i ($i \in I$) with $\sum_{i \in I} x_i \leq 1$ be given. If all x_i are non-positive, then $\sum_{i \in I} f_1(x_i) \leq 0$ is immediately clear. Otherwise, one obtains

$$\sum_{i \in I} f_1(x_i) < \sum_{i \in I} (k+1) \times x_i / k \leq 1 + 1/k$$

and also

$$k \times \sum_{i \in I} f_1(x_i) \leq \sum_{i \in I} \lfloor (k+1) \times x_i \rfloor.$$

Since the right-hand side is an integer with

$$k \times \sum_{i \in I} f_1(x_i) < k + 1,$$

it follows that

$$k \times \sum_{i \in I} f_1(x_i) \leq k.$$

The monotone function f_3 differs from f_1 only in points $x > 1$ where $(k+1)*x \in \mathbb{N}$. One gets for these x that

$$f_3(x) = ((k+1) \times x - 1)/k$$
$$= f_3(x-\varepsilon)$$
$$= f_1(x-\varepsilon)$$
$$> f_1(x).$$

Since f_1 is a general dual-feasible function, and $f_3(x) = \lim_{y \uparrow x} f_1(y)$ for any $x \in \mathbb{R}$, the function f_3 is also a general dual-feasible function. Since f_3 dominates f_1, the function f_1 is not a maximal general dual-feasible function. f_3 is symmetric and can therefore not be dominated by another general dual-feasible function. Hence, f_3 is a maximal general dual-feasible function.

3.8. The given function g is a Hölder continuous classical maximal dual-feasible function, because there is a constant $c > 0$, such that it holds for every $x, y \in [0, 1]$ that

$$|g(x) - g(y)| \leq c \times \sqrt{|x-y|},$$

and g is strict convex on $[0, 1/2]$, symmetric and therefore superadditive, and $g(0) = 0$. Let $p := 1$, $y := 1/2$ and $x \approx 0$ with $x < 0$. Then, $g(y) = 1/2, f(x) = tx$ and $g(x+y) = (1 - \sqrt{-2x})/2$. Hence, we have

$$f(x+y) - f(x) - f(y) = -tx - \sqrt{-x/2} < 0$$

for $x > \frac{-1}{2t^2}$.

Chapter 4

4.1. Check the different classes of VP-MDFF, or show that there are two elements **x** and **y** such that $\mathbf{x} + \mathbf{y} \leq \mathbf{w}$ and $f(\mathbf{x}) + f(\mathbf{y}) > 1$.

4.2. Check the different classes of VP-MDFF.

4.3. Immediate.

4.4. Immediate.

4.5. First, note that applying the function to each item is equivalent to cutting the pieces into squares and applying two maximal dual-feasible functions on the resulting 2-OPP-O instance.

By writing the corresponding values of function, and using the superadditivity of f and g, question 3 is directly answered. Question 4 is similar, and can be answered by induction on w and h and using question 3.

4.6. One can use the same methods used for the m-OPP-R.

4.7. Check that for all feasible patterns, the sum of the images is less than or equal to 1. Note that 1 and 2 are two different elements, although they have the same size.

4.8. The values of $f(1)$ and $f(2)$ can be increased.

4.9. A solution is almost given in the next question. This cannot happen with a CS-MDFF because such a function only depends on the sizes and is increasing.

4.10. Choose J as large as possible, and remark that the presence of item 5 does not modify the optimal solution.

4.11. Choose the node set $\{1, 2, 9, 10\}, \{1, 2, 3, 4\}, \{1, 3, 7, 8\}, \{3, 4, 5, 6, 11, 12\}$.

4.12. The problem is equivalent to the special case of graph colouring, where the graph is an interval graph.

Chapter 5

5.1.

(a) False.
Without superadditivity, counter examples exist like the following one: $1/2x \leq 1$ is given, and hence $x := 2$ is feasible. Suppose $f(1/2) = 1/2$ and $f(1) = 0$. That yields the contradiction $f(1/2) \times 2 \leq f(1)$ or $1 \leq 0$.

(b) False.
A superadditive function f with $f(0) < 0$ may also yield contradictions. For instance, applying such a function to the inequality $x \leq 0$, which allows $x := 0$, could yield the false conclusion $x < 0$.

5.2. This VP-MDFF maps the vectors $\left(\frac{5}{11}, \frac{3}{8}\right)^\top$ and $\left(\frac{4}{11}, \frac{1}{2}\right)^\top$ to $1/2$ and the vector $\left(\frac{3}{11}, \frac{1}{4}\right)^\top$ to $1/4$. That yields the valid inequality

$$\frac{x_1}{2} + \frac{x_2}{2} + \frac{x_3}{4} \leq 1,$$

which is equivalent to the demanded one.

(Remark: The demanded inequality could also be obtained by adding the two inequalities, dividing by 4 and applying the rounding procedure due to Chvátal and Gomory.)

5.3. The definition of the VP-DFF implies $f(\mathbf{a}^{j_0}) + f(\mathbf{a}^j) \leq 1$, and hence $f(\mathbf{a}^j) \leq 0$. Because of the range of f, it follows that $f(\mathbf{a}^j) = 0$. That implies also that if a column vector $\mathbf{a}^{j_0}$ gets the special mapping to 1 due to the property "large argument vector" then every "small argument vector" $\mathbf{a}^j$ with $\mathbf{a}^{j_0} + \mathbf{a}^j \leq \mathbf{w}$ will get the special mapping to 0, and the other VP-MDFF in the considered construction principles will not be applied to these vectors. Moreover, the special mapping to 1 can be applied only to vectors $\mathbf{a}^j$ with $2\mathbf{a}^j \not\leq \mathbf{w}$.

References

Aardal K, Weismantel R (1997) Polyhedral combinatorics. Wiley, New York
Alves C (2005) Cutting and packing: problems, models and exact algorithms. PhD thesis, Universidade do Minho, Guimaraes
Alves C, Valério de Carvalho J, Clautiaux F, Rietz J (2014) Multidimensional dual-feasible functions and fast lower bounds for the vector packing problem. Eur J Oper Res 233:43–63
Bazaraa M, Jarvis J, Sherali H (2010) Linear programming and network flows. Wiley, New York
Boschetti M, Mingozzi A (2003) The two-dimensional finite bin packing problem. Part I: new lower bounds for the oriented case. 4 OR 1:27–42
Burdett C, Johnson E (1977) A subadditive approach to solve linear integer programs. Ann Discret Math 1:117–144
Caprara A, Locatelli M, Monaci M (2005) Bilinear packing by bilinear programming. In: Jünger M, Kaibel V (eds) Integer programming and combinatorial optimization, 11th international IPCO conference, Berlin, 8–10 June 2005. Lecture notes in computer science, vol 3509. Springer, Berlin, pp 377–391
Carlier J, Néron E (2007a) Computing redundant resources for cumulative scheduling problems. Eur J Oper Res 176(3):1452–1463
Carlier J, Néron E (2007b) Computing redundant resources for the resource constrained project scheduling problem. Eur J Oper Res 176(3):1452–1463
Carlier J, Clautiaux F, Moukrim A (2007) New reduction procedures and lower bounds for the two-dimensional bin-packing problem with fixed orientation. Comput Oper Res 34:2223–2250
Chvátal V (1973) Edmonds polytopes and a hierarchy of combinatorial problems. Discret Math 4:305–337
Clautiaux F (2010) New collaborative approaches for bin-packing problems. Habilitation à Diriger des Recherches, Université de Lille 1, France
Clautiaux F, Jouglet A, Hayek J (2007) A new lower bound for the non-oriented two-dimensional bin-packing problem. Oper Res Lett 35:365–373
Clautiaux F, Alves C, Valério de Carvalho J (2010) A survey of dual-feasible and superadditive functions. Ann Oper Res 179:317–342
Dantzig GB, Wolfe P (1960) Decomposition principle for linear programs. Oper Res 8:101–111
Dash S, Günlük O (2006) Valid inequalities based on simple mixed-integer sets. Math Program 105:29–53
Fekete S, Schepers J (2001) New classes of fast lower bounds for bin packing problems. Math Program 91:11–31

Fekete S, Schepers J (2004) A general framework for bounds for higher-dimensional orthogonal packing problems. Math Meth Oper Res 60:311–329

Geoffrion A (1974) Lagrangian relaxation and its uses in integer programming. Math Program Study 2:82–114

Gilmore P, Gomory R (1961) A linear programming approach to the cutting stock problem (part I). Oper Res 9:849–859

Gomory R (1958) Outline of an algorithm for integer solutions to linear programs. Bull Am Math Soc 64:275–278

Johnson D (1973) Near optimal bin packing algorithms. Dissertation, Massachussetts Institute of Technology, Cambridge, MA

Khanafer A, Clautiaux F, Talbi E (2010) New lower bounds for bin packing problems with conflicts. Eur J Oper Res 206:281–288

Letchford A, Lodi A (2002) Strengthening Chvával-Gomory cuts and Gomory fractional cuts. Oper Res Lett 30:74–82

Lueker G (1983) Bin packing with items uniformly distributed over intervals [a,b]. In: Proceedings of the 24th annual symposium on foundations of computer science (FOCS 83). IEEE Computer Society, Silver Spring, MD, pp 289–297

Martello S, Toth P (1990) Knapsack problems - algorithms and computer implementation. Wiley, Chichester

Nemhauser G, Wolsey L (1998) Integer and combinatorial optimization. Wiley, New York

Rietz J, Alves C, Valério de Carvalho J (2010) Theoretical investigations on maximal dual feasible functions. Oper Res Lett 38:174–178

Rietz J, Alves C, Valério de Carvalho J (2012a) On the extremality of maximal dual feasible functions. Oper Res Lett 40:25–30

Rietz J, Alves C, Valério de Carvalho J, Clautiaux F (2012b) Computing valid inequalities for general integer programs using an extension of maximal dual-feasible functions to negative arguments. In: Proceedings of the 1st international conference on operations research and enterprise systems (ICORES 2012)

Rietz J, Alves C, Valério de Carvalho J, Clautiaux F (2014) On the properties of general dual-feasible functions. In: Murgante B, Misra S, Rocha AMAC, Torre C, Rocha JG, Falcão MI, Taniar D, Apduhan BO, Gervasi O (eds) Computational science and its applications – ICCSA 2014. Lecture notes on computer science, vol 8580. Springer, pp 180–194. doi:10.1007/978-3-319-09129-7_14. http://dx.doi.org/10.1007/978-3-319-09129-7_14

Rietz J, Alves C, Valério de Carvalho J, Clautiaux F (2015) Constructing general dual-feasible functions. Oper Res Lett 43:427–431

Robertson N, Seymour P (1986) Graph minors. II algorithmic aspects of tree-width. J Algorithms 7:309–322

Rose D, Tarjan E, Lueker G (1976) Algorithmic aspects of vertex elimination on graphs. SIAM J Comput 5:146–160

Spieksma F (1994) A branch-and-bound algorithm for the two-dimensional vector packing problem. Comput Oper Res 21:19–25

Valério de Carvalho J (1999) Exact solution of bin packing problems using column generation and branch-and-bound. Ann Oper Res 86:629–659

Valério de Carvalho J (2002) A note on branch-and-price algorithms for the one-dimensional cutting stock problems. Comput Optim Appl 21:339–340

Vance P (1998) Branch-and-Price algorithms for the one-dimensional cutting stock problem. Comput Optim Appl 9:211–228

Vanderbeck F (2000) Exact algorithm for minimizing the number of setups in the one-dimensional cutting stock problem. Oper Res 46(6):915–926

Index

Bin-packing problem (BP), 113
Bin-packing problem with conflicts (BPC), 115
Bin-packing with conflicts DFF, 115
Binary knapsack problem with conflicts (KPC), 115
Binary knapsack problem, KP–01, 114
Block angular structure, 8

Column generation, 10
   master problem, 10
   subproblem, 10
Composition, 34, 53, 66
Cutting Stock Problem, 7, 23

Dantzig-Wolfe decomposition, 3
Data-dependent set-covering dual-feasible function (SC-DDFF), 93
Discrete dual-feasible function, 22
Dual-feasible function, 21

Extremality, 28, 56

General dual-feasible function, 52
Gilmore and Gomory model, 7, 23

Integrality constraints, 1

Integrality property, 4

LP relaxation, 1

Maximality, 25, 54
m-dimensional orthogonal bin-packing problem (m-OPP), 111
Minkowski's theorem, 3
Multi-dimensional knapsack, mD-KP, 97

Orthogonal packing DFF (m-OPP-R-DFF, m-OPP-O-DFF), 111

Set-covering dual-feasible function (SC-DFF), 94
Superadditivity, 25
Symmetry, 26, 35

Tree-decomposition (of a graph), 117
Triangulated (graph), 117

Vector packing (mD-VPP), 95
Vector packing dual-feasible function (VP-DFF), 97

MIX
Papier aus verantwortungsvollen Quellen
Paper from responsible sources
FSC® C105338

If you have any concerns about our products,
you can contact us on
ProductSafety@springernature.com

In case Publisher is established outside the EU,
the EU authorized representative is:
**Springer Nature Customer Service Center GmbH
Europaplatz 3, 69115 Heidelberg, Germany**

Printed by Libri Plureos GmbH
in Hamburg, Germany